Event-Triggered Identification of Systems with Quantized Observations

Guo Jin Diao Jingdong

Beijing
Metallurgical Industry Press
2020

Abstract

There are 8 chapters in this book. Chapter 1 introduces the system identification problem with quantized and event-triggered observations. Chapter 2 gives a brief review on several optimization and identifcaion techniques. Chapters 3 ~ 6 focus on the linear system identifaicon under the send-on-delta mechanism (predetermined and adaptive), the prediction-based communication scheme and the either-or communication scheme. Chapters 7 ~ 8 turn the attention to two kinds of basic block-oriented nonlinear systems—Wiener systems and Hammerstein systems.

The book is written for researchers and engineers working in systems and control, communication and computer networks, etc.

图书在版编目(CIP)数据

量化系统的事件驱动辨识=Event-Triggered Identification of Systems with Quantized Observations：英文/郭金，刁靖东编著．—北京：冶金工业出版社，2019.1（2020.1 重印）

ISBN 978-7-5024-7993-0

Ⅰ.①量… Ⅱ.①郭… ②刁… Ⅲ.①系统辨识—英文 Ⅳ.①N945.14

中国版本图书馆 CIP 数据核字（2018）第 301665 号

出 版 人　陈玉千
地　　址　北京市东城区嵩祝院北巷 39 号　邮编　100009　电话　(010)64027926
网　　址　www.cnmip.com.cn　电子信箱　yjcbs@cnmip.com.cn
责任编辑　戈　兰　美术编辑　彭子赫　版式设计　孙跃红
责任校对　石　静　责任印制　李玉山
ISBN 978-7-5024-7993-0
冶金工业出版社出版发行；各地新华书店经销；北京虎彩文化传播有限公司印刷
2019 年 1 月第 1 版，2020 年 1 月第 3 次印刷
169mm×239mm；9.5 印张；182 千字；138 页
64.00 元

冶金工业出版社　投稿电话　(010)64027932　投稿信箱　tougao@cnmip.com.cn
冶金工业出版社营销中心　电话　(010)64044283　传真　(010)64027893
冶金工业出版社天猫旗舰店　yjgycbs.tmall.com
（本书如有印装质量问题，本社营销中心负责退换）

Preface

In current era of information processing where thousands, even millions, of computers are interconnected using a heterogeneous network of wireless and wired systems, how to save the communication resource has become a very important and interesting problem. The quantization and the event-triggered communication scheme is widely utilized in practice. The former is to divide the data range into a finite collection of subsets, and usually a process from a large set to a smaller set. The latter designs an "event" to determine whether the information is "important enough" or not. Only when the event occurs, the information will be sent, and hence the number of times of transmission can be down.

This book presents new methodologies that utilize quantized and event-triggered information in system identification. From the viewpoint of targeted plants, it treats both linear and nonlinear systems with the binary-valued quantization and the multi-level quantization. Several event-triggered communication schemes are designed, including the send-on-delta mechanism (predetermined and adaptive), the prediction-based communication scheme and the either-or communication scheme. The key methodologies of the book combine the empirical measure, the weighted least squares optimization, the stochastic approximation technique to cover convergence, convergence rate, estimator efficiency, communication rate and the tradeoff between the estimation performance and the communication cost.

The book is written for researchers and engineers working in systems and control, communication and computer networks, signal processing, remote sensing, and mathematical statistics, etc. Selected materials from the book may also be used in a graduate-level course on system identification.

This book could not have been completed without the help and encouragement of many people. We recognize our institutions and colleagues for providing us with a stimulating and supportive academic environment. We are deeply indebted to many researchers in the field for insightful discussions and constructive criticisms, and for enriching us with their expertise and enthusiasm. This book is supported by the National Natural Science Foundation of China under Grant 61773054.

Guo Jin　Diao Jingdong

Beijing, September 2018

Contents

Notations

$\mathbf{R}$	set of real numbers
$\mathbf{R}^n$	n-dimensional real-valued Euclidean space
$\mathbf{R}^{m\times n}$	set of real-valued $m\times n$ matrixes
$\varnothing$	empty set
$\mathbf{N}_+$	set of positive integers
L	$\{1, \cdots, l\}$
L_0	$\{1, \cdots, l_0\}$
L_0^-	$\{l_0+1, \cdots, l\}$
$\in$	belong to
$\notin$	not belong to
$\lvert\cdot\rvert$	absolute value of a scalar
I_A	indicator function of the set A
$\lfloor z \rfloor$	floor function: the largest integer that is $\leqslant z$
$\lceil z \rceil$	ceil function: the least integer greater than or equal to z
$\log x$	logarithm of x
arg min	argument of the minimum
$\max\{a_1, \cdots, a_n\}$	maximum value among $a_1, \cdots, a_n$
$\min\{a_1, \cdots, a_n\}$	minimum value among $a_1, \cdots, a_n$
lim	limit of a sequence
lim sup	superior limit
$\xrightarrow{d}$	convergence in distribution
sup	supremum
$O(y)$	function of y satisfying $\sup_{y\neq 0} \lvert O(y)\rvert / \lvert y\rvert < \infty$
$o(y)$	function of y satisfying $\lim_{y\to 0} o(y) / \lvert y \rvert = 0$
$\mathbf{1}$	column vector with all elements equal to one

I	identity matrix of suitable dimension
$\mathrm{diag}(a_1, \cdots, a_n)$	diagonal matrix of $a_1, \cdots, a_n$
$\boldsymbol{A}^{\mathrm{T}}$	transpose of a vector or a matrix
$\boldsymbol{A}^{-1}$	inverse of a matrix $\boldsymbol{A}$
$\boldsymbol{A}^{+}$	Moore-Penrose inverse of a matrix $\boldsymbol{A}$
$\\|\cdot\\|$	Euclidean norm of a vector or the Frobenius norm of a matrix
$\Pr(\cdot)$	probability measure
$F(\cdot)$	cumulative distribution function
$f(\cdot)$	density function $f(x) = \mathrm{d}F(x)/\mathrm{d}x$
$\mathcal{F}$	σ-algebra
$(\Omega, \mathcal{F}, \Pr)$	probability space
$(\Omega, \mathcal{F}, \{\mathcal{F}_t\}, \Pr)$	filtered probability space
EX	expectation of a random variable X
$E\{X \mid \mathcal{F}\}$	conditional expectation of X given $\mathcal{F}$
$\{\mathcal{F}_t\}$	filtration $\{\mathcal{F}_t, t \geqslant 0\}$
$\mathcal{N}(\mu, \sigma^2)$	normal distribution with mean μ and variance σ^2
i. i. d.	independent and identically distributed
MDS	martingale difference sequence
w. p. 1	with probability one
a. e.	almost everywhere
a. s.	almost sure
CRLB	Cramér-Rao lower bound
SISO	single-input-single-output
FIR	finite impulse response
EC	estimator center
ϕ_k	regression (column) vector of θ at time k
θ	system parameter (column) vector
Θ	set of possible values of θ
C	threshold of binary-valued quantized sensor
$C_1, \cdots, C_m$	thresholds of multi-threshold quantized sensor

QCCE	quasi-convex combination estimate
$\hat{\theta}_k$	estimation of θ at time k
$\tilde{\theta}_k$	estimation error, that is, $\tilde{\theta}_k = \hat{\theta}_k - \theta$
γ_k	event-triggering indicator
$\bar{\gamma}_k$	average communication indicator
$\bar{\gamma}$	communication rate
s. t.	subject to
$:=$	defined as
$\square$	end of a proof

1 Introduction

1.1 Motivation

Identification, state estimation and control theory are three interpenetrating fields of modern control systems. Identification and state estimation relies on the support of control theory, and control theory can hardly be applied without identification and state estimation technology. One of the keys to practical application of control theory is fully understanding the motion law of the study object and establishing the corresponding mathematical model when representing their causality relationship. Since the term "identification" was first introduced by American scholar Zadeh in the well-known paper[1] and 1950s, system identification has been developed as a modeling method for the control design of dynamic systems for more than half a century, and has achieved rich theoretical results. Its practical application is also across nearly all fields and industries[2].

In terms of conventional control problems, system identification has been a mature research field. Its ideas and methods continually support the development of relevant theories, while the study of system identification has never stopped. With the advent of the information technology revolution, the fields of industrial engineering, aerospace and others that automation technology facing directly have undergone tremendous changes. Traditional industrial engineering mostly focuses on the modeling and control of a single device, but now it often faces groups interrelated in both time and space, such as sensor networks, multi-agent systems and smart power grids with new energy. A development trend of physical systems is that systems are increasingly interconnected, and many are interconnected by communication networks. Traditional control processes usually control only one process at a time, but today, unlike these traditional systems, there are a lot of interactions and adjustments within systems. Therefore, how to save communication resources between systems has become an important issue that we have to face.

In such situations, the thought of "event-trigger" arises as the times require. It means that the acquisition of signals is driven by a specific event, rather than by time. It can reduce the data communication rate, meanwhile, and ensure the control quality. The property and type of the event depends on the actual goal, and it is designed to collect the data with "maximum amount of information". It can be defined as a variable exceeding a specified value, or a packet reaching a specified node, and so on. The thought of "event-trigger" has attracted wide attention in control field since it was put forward, and a series of research achievements have been made in the design of controller, state estimation and so on[3~12].

In fact, the event-triggered mechanism has many applications in practical industrial. For example, in the tank water level control, the "event" can be set as the water level exceeds a certain limit value. When the system runs, the water level sensor transmits the signal to a chip to detect the event through the signal transmission system, and finally decides whether to drive the controller to control the water injection and drainage of the tank[13]. In human bionic machines, their humanoid action is also based on events rather than time. Similarly, when an industrial process object is operated manually by an engineer, the corresponding operation is judged by events, not by time-driven. A new instruction will be given by the operator only when the value of the controlled signal exceeds the set value[14].

Different from traditional methods, which broadcast information continuously or periodically, event-triggered methods update the data according to some real-time performance indicators (triggering conditions) (see Figure 1.1). Generally speaking, the event-triggered sampling cycle is time variant rather than fixed. Intuitively, the fewer events occurring will cause the fewer times of sampling, the smaller calculation load, communication load, energy consumption and system cost, the life of the equipment will also be extended[15]. In [3], the famous scholar K. J. Aström calls the traditional periodic sampling as Riemann sampling and the event-triggered sampling as Lebesgue sampling. An example is given to show the advantage of Lebesgue sampling over Riemann sampling, which is similar to the advantage of Lebesgue integral over Riemann integral to some extent. A large number of research has shown that event-triggered mechanism has prominent advantages in saving network energy consumption, and can effectively reduce the communication load as well. However, it is relatively complex and difficult to analyze system performances and design identification algorithms.

Event-triggered mechanism reduces the data communication rate while guaranteeing the system performance, but how to save the power and the bandwidth of signal transmis-

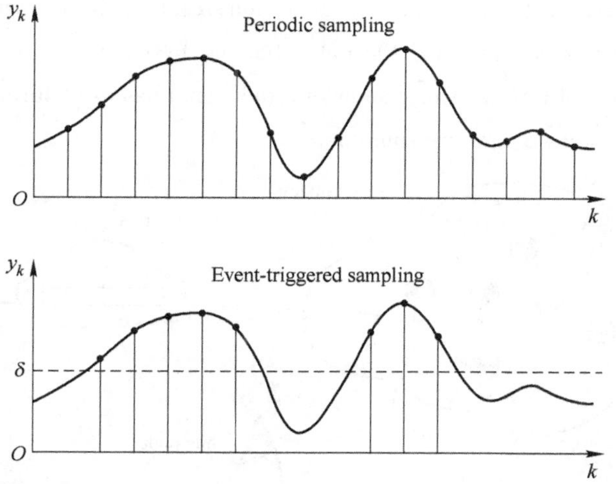

Fig. 1.1 Periodical sampling versus event-triggered sampling (the triggering condition is $y_k > \delta$)

sion, prevent network congestion and ensure the reliability of communication is another problem that we must face. Typical communication systems include processes of sampling, quantization, data compression and coding for reducing data size and increasing transmission reliability. In the process of transmission, packet loss, random transmission delay and erroneous received data are very common phenomena. These new uncertainties bring challenges to system identification, and identification with quantized observations is one of the latest research directions in this field[16~22].

In today's cyber age of pervasive communication and information technology, quantization system has been widely utilized in practical production and life, especially in biomedical, digital signal transmission, intelligent sensor networks and other fields which are closely related to modernization[23,24]. For example, as a strategic emerging industry in the 12th Five-Year Plan of China, Internet of Things has an extensive development space. One of the key technologies is wireless sensor networks (Figure 1.2), in which the number of sensors is huge, possibly reaches tens of millions, or even more[25]. However, all these sensors use a common channel when transmit data. Therefore, the bandwidth allotted for each sensor to transmit data is very small. As we all know, in the process of data transmission, the smaller step of quantization causes the larger amount of data, and data transmission will be limited by the bandwidth of the channel. For a bandwidth, too much data will cause channel congestion, resulting in a great delay and even data distortion. In this case, even if the data obtained by each sensor is very accurate, due to the influence of transmission bandwidth, the signal has to be further quantized into very low precision data before transmission.

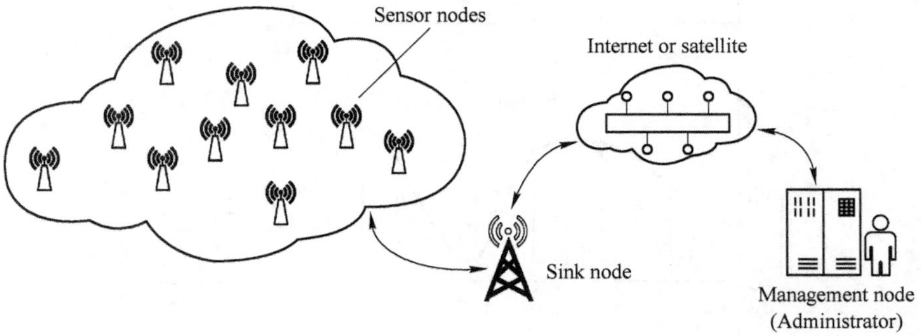

Fig. 1. 2 Wireless sensor network

Quantized measurements contain less information, and the measured signals are non-one-to-one mapped to the input, state and controlled output of the system, the relationship is essentially nonlinearity. Identification and control methods, such as least squares method and maximum likelihood estimation, Kalman filter and self-tuning regulator, which have been developed for traditional linear and nonlinear systems, are no longer applicable to quantized systems. Therefore, it is necessary to develop new methodologies in identification, controller design and comprehensive analysis for quantized data[26~31].

On the whole, event-triggered mechanism can effectively reduce the data communication rate under the premise of guaranteeing the control quality, while quantization mechanism can save the communication bandwidth and storage. The former breaks through the traditional periodic sampling in term of time, sampling only at those moments when "the amount of information reaches a certain degree of richness", and the latter, in term of space, saves bandwidth for transmission and capacity for data storage. The joint use of the two mechanisms is an inevitable choice for communication between systems. In the face of this new way of data acquisition, control theory is bound to be developed and improved. The event triggered system identification with quantized observations is proposed under such circumstances. This research direction is the development of conventional system identification in the new era and the inevitable trend of network and information technology. It involves control theory, communication science, computer science and other fields, oriented to systems interconnected in time and space, aims to solve the modeling problem under the new way of data acquisition. Inevitably, the relevant results will be applied to many fields, such as Cyber-Physical systems, sensor networks, multi-agent systems, smart power grids and so on[32~36].

1.2 Event-Triggered Identification with Quantized Observations

Consider a system:

$$y_k = \mathbb{G}(u_k, \theta) + d_k$$

where u_k is the input, d_k is the system noise, θ is the unknown parameter vector to be identified.

The output y_k is measured by a sensor of m thresholds $-\infty < C_1 < \cdots < C_m < \infty$, which can be represented by as et of m indicator functions

$$s_k = [s_k^1, \cdots, s_k^m]^{\mathrm{T}}$$

where

$$s_k^i = I_{\{-\infty < y_k \leqslant C_i\}}, \quad i = 1, \cdots, m \tag{1.1}$$

An alternative representation of (1.1) is by defining

$$\tilde{s}_k = \sum_{i=1}^{m+1} i \, \tilde{s}_k^l$$

where

$$\tilde{s}_k^i = I_{\{C_{i-1} < y_k \leqslant C_i\}}, \quad i = 1, \cdots, m+1$$

as (a) of Figure 1.3. Hence, $\tilde{s}_k = i$, for $i = 1, \cdots, m+1$, implies that $y_k \in (C_{i-1}, C_i]$ with $C_0 = -\infty$ and $C_{m+1} = \infty$. The simplest and basic quantized form is the binary-valued observation, that is, $m = 1$ and only one threshold C in the quantized sensor, which can be given by

$$s_k = I_{\{y_k \leqslant C\}}$$

as (b) of Figure 1.3.

To reduce the number of transmission, a local event detector is used to decide

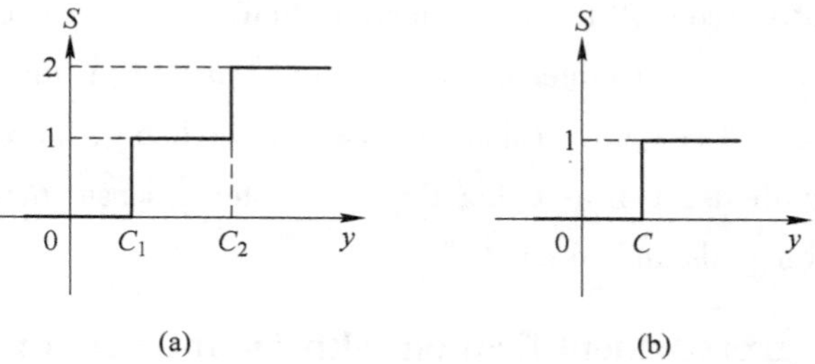

Fig. 1. 3　Quantized observations

(a) Multi-threshold quantized observations; (b) Binary-valued observations

whether the quantized observation s_k is transmitted to the remote EC (Estimation Center), as shown in Figure 1. 4. Let γ_k denote the transmission indicator, i. e., $\gamma_k = 1$ indicates that the detector is triggered and the sensor data s_k is send to the EC at time k, while $\gamma_k = 0$ means the communication is denied. γ_k can be predetermined or computed in real time according to the past information. In general, γ_k is a function of $u_1, \cdots, u_{k-1}, s_1, \cdots, s_{k-1}$. A frequently encountered case is that γ_k is a function of s_k and $\hat{\theta}_{k-1}$, where $\hat{\theta}_{k-1}$ denotes the estimate of θ at time $k-1$, and is either broadcasted by the estimator center or computed by the event detector itself. For evaluating the capacity of the event-triggered mechanism in saving the communication resources, we define the communication rate as

$$\bar{\gamma} = \lim_{k\to\infty} \frac{1}{k} \sum_{i=1}^{k} \gamma_i \tag{1.2}$$

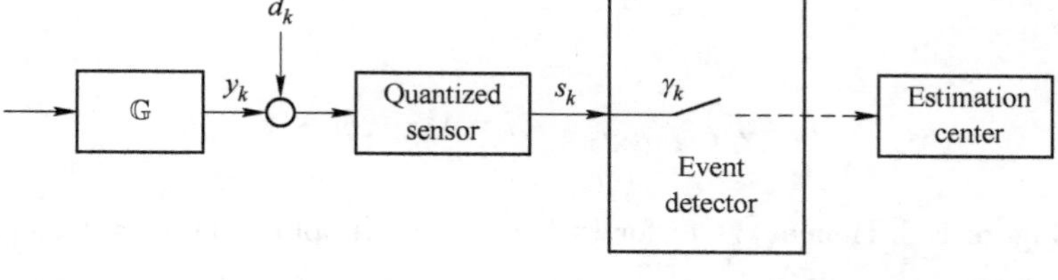

Fig. 1. 4　Event-triggered identification with quantized observations

For the EC, the available information is $\{\gamma_k, \gamma_k s_k\}$. Under such an estimation framework called *event-triggered identification with quantized observations*, some fundamental issues emerge.

(1) How to utilize $\{\gamma_k, \gamma_k s_k\}$ to construct identification algorithms? How to develop new methodologies for establishing the convergence performance, including the convergence, the convergence rate and the asymptotic efficiency in terms of the CRLB (Cramér-Rao lower bound)?

(2) How to resolve the tradeoff between the estimation quantity and the communication rate? To be more specific, how to design γ_k for the minimum communication rate under an allowable estimation error or how to design γ_k for the minimum estimation error under an allowable communication rate?

This book aims to answer the questions above from view of system models, quantization schemes, event-triggered mechanisms, and so on.

1.3 Outline of the Book

Roughly speaking, this book is concerned with how the quantization and the event-triggered communication scheme jointly affect the identification algorithm and the convergence performance. The relationship among chapters is illustrated by Table 1.1. Chapter 3 ~ 6 focus on the linear time-invariant stable discrete-time system. Chapter 7 ~ 8 turn the attentions to two kinds of basic block-oriented nonlinear systems—Wiener systems and Hammerstein systems. Below, each of these chapters is briefly described and commented.

Table 1.1 Chapter overview

<table>
<tr><th colspan="2">Plant</th><th>Chapter No.</th><th>Event-Triggered Mechanism</th><th>Quantization Scheme</th></tr>
<tr><td colspan="2" rowspan="4">Linear System</td><td>3</td><td>Scheduling</td><td rowspan="2">Binary-Valued Quantization</td></tr>
<tr><td>4</td><td>General Send-on-Delta Event-trigger</td></tr>
<tr><td>5</td><td>Prediction-Based Event-trigger</td><td rowspan="2">Multi-Threshold Quantization</td></tr>
<tr><td>6</td><td>Either-or Communication Scheme</td></tr>
<tr><td rowspan="2">Nonlinear System</td><td>Wiener System</td><td>7</td><td>Either-or Communication Scheme</td><td>Binary-Valued Quantization</td></tr>
<tr><td>Hammerstein System</td><td>8</td><td>Prediction-Based Event-trigger</td><td>Multi-Threshold Quantization</td></tr>
</table>

Chapter 2 gives a brief review on the least square optimization, the stochastic approximation technique, the empirical-measure-based identification and the quantized input-output identification. They can lay a foundation for the subsequent chapters.

Chapter 3 focuses on FIR (finite impulse response) systems to study the identification with scheduled binary-valued observations. For the convenience of discussion, we proceed on the periodic input. An empirical-measure-based identification algorithm is introduced and its convergence performance is given. The communication rate is derived and the input design is studied for the minimum communication rate.

Chapter 4 pays attention to the adaptive send-on-delta communication mechanism, that is, the event-trigger depends on the past information and is adjusted in real time. For the binary-valued observations, we take the stochastic approximation technique to design recursive identification algorithms. Under sufficiently rich inputs, the convergence and convergence rate is obtained. The event-triggering conditions are given to ensure the convergence properties of the algorithm. The communication rate is discussed.

Chapter 5 proposes a predication-based communication scheme, where the difference between the quantized observation and its prediction is used to construct triggering event. For the binary-valued output quantization, by full use of the received data, the triggered indicator, the quantization threshold, and the statistical property of the system noise, an identification algorithm is proposed to estimate the unknown parameter based on the empirical-measure-based identification technique and the weighted least-squares optimization. Under quantized inputs, the convergence performance is established and the communication is given. The tradeoff between the communication cost and the estimation performance is formulated to a constrained minimization problem and discussed. The method and results are extended to the case of multi-threshold quantized observations.

Chapter 6 presents a so-called either-or communication scheme by considering the particularity of the binary-valued quantization and the statistical property of the system noise. The multi-threshold quantization can be seen as a vector-valued observation in which each component represents the output of one threshold, and then the either-or communication scheme can be applied to each component of the multi-threshold quantization. For FIR systems, the corresponding identification algorithm, the convergence properties and the communication rate are given.

Chapter 7 identifies the quantized-input Wiener systems under the either-or communication scheme and binary-valued output observations. The identifiability problem is discussed first. Then a three-step identification algorithm is proposed to estimate the unknown parameter for the identifiable systems. Under some typical assumptions on system order, input persistent excitation, and noise distribution functions, the algorithm is shown to be strongly convergent and asymptotically efficient in terms of the CRLB. The communication rate is given.

Chapter 8 is the identification of discrete-time quantized-input Hammerstein systems with the prediction-based communication scheme and quantized output observations. Based on the modified empirical measure and the quasi-convex combination technique, identification algorithms are given and their convergence properties are derived, together with the communication rate.

2 Preliminaries

2.1 Least Square

The method of least squares is a standard approach in regression analysis to approximate the solution of overdetermined systems, i. e. , sets of equations in which there are more equations than unknowns. "Least squares" means that the overall solution minimizes the sum of the squares of the residuals made in the results of every single equation.

The objective consists of adjusting the parameters of a model function to best fit a data set. A simple data set consists of N points (data pairs) $(x_i, y_i), i = 1, \cdots, N$, where x_i is an independent variable and y_i is a dependent variable whose value is found by observation. The model function has the form $g(x, \theta)$, where m adjustable parameters are held in the vector θ. The goal is to find the parameter values for the model that "best" fits the data. The fit of a model to a data point is measured by its residual, defined as the difference between the actual value of the dependent variable and the value predicted by the model:

$$r_i = y_i - g(x_i, \theta)$$

The least-squares method finds the optimal parameter values by minimizing the sum, S, of squared residuals:

$$S = \sum_{i=1}^{N} r_i^2$$

The minimum of the sum of squares is found by setting the gradient to zero. Since the model contains m parameters, there are m gradient equations:

$$\frac{\partial S}{\partial \theta_j} = 2 \sum_i r_i \frac{\partial r_i}{\partial \theta_j} = 0, \quad j = 1, \cdots, m$$

and since $r_i = y_i - g(x_i, \theta)$, the gradient equations become

$$-2\sum_{i} r_i \frac{\partial g(x_i,\theta)}{\partial \theta_j} = 0, \quad j = 1,\cdots,m$$

The gradient equations apply to all least squares problems. Each particular problem requires particular expressions for the model and its partial derivatives.

A regression model is a linear one when the model comprises a linear combination of the parameters, i. e. ,

$$g(x,\theta) = \sum_{j=1}^{m} \theta_j \phi_j(x)$$

where the function ϕ_j is a function of x . Letting $X_{ij} = \phi_j(x_i)$ and $Y = (y_i)$, we can then see that in that case the least square estimate (or estimator, in the context of a random sample), θ is given by

$$\hat{\theta} = (X^{\mathrm{T}}X)^{-1}X^{\mathrm{T}}Y$$

In some cases the observations may be weighted—for example, they may not be equally reliable. In this case, one can minimize the weighted sum of squares:

$$S = \sum_{i=1}^{N} w_i r_i^2$$

where w_i is the weight of the ith observation. This is called the Weighted Least Square. The estimated parameter values are linear combinations of the observed values

$$\hat{\theta} = (X^{\mathrm{T}}WX)^{-1}X^{\mathrm{T}}WY$$

where $W = \mathrm{diag}(w_i)$ is the diagonal matrix of the weights. The weights should, ideally, be equal to the reciprocal of the variance of the measurement. (This implies that the observations are uncorrelated. If the observations are correlated, the expression $S = \sum_{k}\sum_{j} r_k W_{kj} r_j$ applies. In this case the weight matrix should ideally be equal to the inverse of the variance-covariance matrix of the observations).

2.2 Stochastic Approximation

Stochastic approximation algorithms are recursive update rules that can be used, among other things, to solve optimization problems and fixed point equations (including standard linear systems) when the collected data is subject to noise. In engineering, optimization problems are often of this type, when you do not have a mathematical model of the system (which can be too complex) but still would like to optimize its behavior by adjusting certain parameters.

Mathematically, the goal of these algorithms is to understand properties of a function

$$g(\theta) = E_X[G(\theta, X)]$$

which is the expected value of a function depending on a random variable X, but to do so without evaluating g directly. Instead, the algorithms use random samples of $G(\theta, X)$ to efficiently approximate properties of g such as zeros or extrema.

The Robbins-Monro algorithm, introduced in 1951 by Herbert Robbins and Sutton Monro, presented a methodology for solving a root finding problem, where the function is represented as an expected value. Assume that we have a function $M(\theta)$ and a constant α, such that the equation $M(\theta) = \alpha$ has a unique root at θ^*. It is assumed that while we cannot directly observe the function $M(\theta)$, we can instead obtain measurements of the random variable $N(\theta)$ where $E[N(\theta)] = M(\theta)$. The structure of the algorithm is to then generate iterates of the form:

$$\theta_{n+1} = \theta_n - a_n(N(\theta_n) - \alpha)$$

Here, $a_1, a_2, \cdots$ is a sequence of positive step sizes. Robbins and Monro proved that θ_n converges in L_2 (and hence also in probability) to θ, and Blum later proved the convergence is actually with probability one, provided that:

(1) $N(\theta)$ is uniformly bounded;

(2) $M(\theta)$ is nondecreasing;

(3) $\left.\dfrac{\mathrm{d}M(\theta)}{\mathrm{d}\theta}\right|_{\theta=\theta^*}$ exists and is positive;

(4) The sequence a_n satisfies the following requirements:

$$\sum_{n=0}^{\infty} a_n = \infty \quad \text{and} \quad \sum_{n=0}^{\infty} a_n^2 < \infty$$

A particular sequence of steps which satisfy these conditions, and was suggested by Robbins-Monro, have the form: $a_n = a/n$, for $a > 0$. Other series are possible but in order to average out the noise in $N(\theta)$, the above condition must be met.

There have been a large amount of applications to practical problems and research works on theoretical issues of the stochastic approximation. See [37] and [38] for more details.

2. 3 Empirical-Measure-Based Identification

Consider a gain system: $y_k = u_k\theta + d_k$. Choose u_k to be a constant. Without loss of generality, assume $u_k \equiv 1$. Then

$$y_k = \theta + d_k \tag{2.1}$$

The output y_k is measured by a sensor of m thresholds $-\infty < C_1 < \cdots < C_m < \infty$, which can be represented by a set of m indicator functions $s_k = [s_k^1, \cdots, s_k^m]^T$.

Assumption 2. 1 Suppose that $\{d_k\}$ is a sequence of i. i. d. (independent and identically distributed) random variables. The accumulative distribution function $F(\cdot)$ of d_1 is invertible and the inverse function is twice continuously differentiable. The moment generating function of d_1 exists.

For convenience, we denote

$$\varrho(x) = \frac{F(x)(1-F(x))}{f^2(x)}, \quad \text{for } x \in \mathbf{R} \tag{2.2}$$

$F_i(x) = F(C_i - x)$ and define the matrix functions

$$U(x) = \operatorname{diag}\left[\left(\frac{\partial F_1(x)}{\partial x}\right)^{-1}, \cdots, \left(\frac{\partial F_m(x)}{\partial x}\right)^{-1}\right] \tag{2.3}$$

and

$$P(x) = \begin{bmatrix} F_1(x) - F_1^2(x) & F_1(x) - F_1(x)F_2(x) & \cdots & F_1(x) - F_1(x)F_m(x) \\ F_1(x) - F_1(x)F_2(x) & F_2(x) - F_2^2(x) & \cdots & F_2(x) - F_2(x)F_m(x) \\ \vdots & \vdots & & \vdots \\ F_1(x) - F_m(x)F_1(x) & F_2(x) - F_m(x)F_2(x) & \cdots & F_m(x) - F_m^2(x) \end{bmatrix} \tag{2.4}$$

for any $x \in \mathbf{R}$, where diag $[\cdots]$ represents a diagonal matrix.

Under Assumption 2. 1, $\{y_k\}$ is an i. i. d. sequence that has the accumulative distribution function $F(\cdot - \theta)$. For the system in (2.1), the probability of $\{s_k^i = 1\}$ is

$$p_i = \Pr\{-\infty < y_k \leqslant C_i\} = F(C_i - \theta) = F_i(\theta) \tag{2.5}$$

We begin with estimation of p_i in (2.5). Take N measurements on s_k. Then for $i \in \{1,\cdots,m\}$

$$\xi_N^i = \frac{1}{N}\sum_{k=1}^{N} s_k^i$$

is the sample relative frequency of y_k taking values in $(-\infty, C_i]$ and an unbiased estimator of p_i for each N, i. e., $E\xi_N^i = p_i$.

An estimator θ_N^i of θ can be derived from $\xi_N^i = F_i(\theta_N^i)$. Consequently, $\theta_N^i = F_i^{-1}(\xi_N^i)$ is an estimator for θ. And the θ_N^i, $i = 1,\cdots,m$, are m asymptotically unbiased estimators of θ based on samples of size N. Denote the estimation errors by $\boldsymbol{e}_N = [e_N^1,\cdots,e_N^m]^{\mathrm{T}}$ with $e_N^i = \theta_N^i - \theta$, also use the notation

$$\boldsymbol{\theta}_N = [\theta_N^1,\cdots,\theta_N^m]^{\mathrm{T}}$$

$$\boldsymbol{1} = [1,1,\cdots,1]^{\mathrm{T}}$$

of compatible dimension. It is readily seen that $\boldsymbol{e}_N = \boldsymbol{\theta}_N - \theta\boldsymbol{1}$. Define

$$\boldsymbol{V}_N(\theta) = E\boldsymbol{e}_N\boldsymbol{e}_N^{\mathrm{T}}$$

Since $E\boldsymbol{e}_N \to 0$ as $N \to \infty$, $\boldsymbol{V}_N(\theta)$ is a covariance matrix of $\boldsymbol{e}_N$, and is positive semidefinite.

Define $\boldsymbol{\nu} = [\nu_1,\cdots,\nu_m]^{\mathrm{T}}$ such that $\nu_1 + \cdots + \nu_m = 1$. One can construct an estimator of θ by

$$\hat{\boldsymbol{\theta}}_N = \sum_{i=1}^{m} \nu_l \theta_N^i = \boldsymbol{\nu}^{\mathrm{T}}\boldsymbol{\theta}_N$$

$\hat{\boldsymbol{\theta}}_N$ is called a Quasi-Convex Combination Estimator (QCCE). The term "quasi-convex" is used since ν_i need not be nonnegative. Since θ_N^i is asymptotically unbiased,

$$E\hat{\theta}_N = \boldsymbol{\nu}^{\mathrm{T}}E\boldsymbol{\theta}_N \to \boldsymbol{\nu}^{\mathrm{T}}\theta\boldsymbol{1} = \theta \quad \text{as} \quad N \to \infty$$

Hence, $\hat{\theta}_N$ is an asymptotically unbiased estimator of θ. Moreover, the variance of the estimation error $\hat{\theta}_N - \theta$ is given by

$$\overline{\mathbb{V}}_N^2 := E(\boldsymbol{\nu}^{\mathrm{T}}\boldsymbol{\theta}_N - \theta)^2 = E(\boldsymbol{\nu}^{\mathrm{T}}\boldsymbol{\theta}_N - \boldsymbol{\nu}^{\mathrm{T}}\theta\boldsymbol{1})^2$$

$$= \boldsymbol{\nu}^{\mathrm{T}}Ee_N\boldsymbol{e}_N^{\mathrm{T}}\boldsymbol{\nu} = \boldsymbol{\nu}^{\mathrm{T}}V_N(\theta)\boldsymbol{\nu}$$

That is, the variance is a quadratic form with respect to the vector ν.

The optimal QCCE minimizes $\overline{\mathbb{V}}_N^2$ which is obtained from

$$\mathbb{V}_N^2 = \min_{\nu,\nu^{\mathrm{T}}\boldsymbol{1}=1}\overline{\mathbb{V}}_N^2 = \min_{\nu,\nu^{\mathrm{T}}\boldsymbol{1}=1}\nu^{\mathrm{T}}V_N(\theta)\nu$$

Theorem 2.1 If Assumption 2.1 holds, then the QCCE $\hat{\theta}_N$ converges strongly to θ, i.e., $\hat{\theta}_N \to \theta$ w.p. 1, and has the strong and mean-square convergence rates

$$\hat{\theta}_N - \theta = O(N^{-\frac{1}{2}}\sqrt{\log\log N}) \quad \text{w. p. 1}$$

and

$$N\overline{\mathbb{V}}_N^2 \to \nu^{\mathrm{T}}U(\theta)P(\theta)U(\theta)\nu \quad \text{as} \quad N \to \infty$$

where $U(\cdot)$ and $P(\cdot)$ are given by (2.3) and (2.4). Also the centered and scaled sequence of $\hat{\theta}_N$ is asymptotically normal in the sense that

$$\sqrt{N}(\hat{\theta}_N - \theta) \xrightarrow{d} \mathcal{N}(0, \nu^{\mathrm{T}}U(\theta)P(\theta)U(\theta)\nu)$$

where $\xrightarrow{d}$ denotes convergence in distribution.

Theorem 2.2 Suppose that Assumption 2.1 holds and $V_N(\theta)$ is positive definite. Then the optimal QCCE can be obtained by choosing

$$\boldsymbol{\nu}^* = \frac{\boldsymbol{V}_N^{-1}(\theta)\boldsymbol{1}}{\boldsymbol{1}^{\mathrm{T}}\boldsymbol{V}_N^{-1}(\theta)\boldsymbol{1}}, \quad \hat{\boldsymbol{\theta}}_N = (\boldsymbol{\nu}^*)^{\mathrm{T}}\boldsymbol{\theta}_N$$

and the minimal variance is $\mathbb{V}_N^2 = \dfrac{1}{\boldsymbol{1}^{\mathrm{T}}\boldsymbol{V}_N^{-1}(\theta)\boldsymbol{1}}$.

Theorem 2.3 The CRLB for estimating θ based on observations of $\{s_k\}$ is

$$\mathbb{V}_{CR}^2(N,m) = \left(N\sum_{i=1}^{m+1}\frac{\overline{p}_i^2}{\tilde{p}_i}\right)^{-1}$$

where $\tilde{p}_i = F_i(\theta) - F_{i-1}(\theta)$ and $\bar{p}_i = \partial\tilde{p}_i/\partial\theta$, $i = 1,\cdots,m+1$.

Theorem 2.4 The optimal QCCE is asymptotically efficient in the sense that

$$N\mathbb{V}_N^2 - N\mathbb{V}_{CR}^2(N,m) \to 0 \quad \text{as} \quad N \to \infty$$

2.4 Quantized Input-Output Identification

Consider an SISO (single-input-single-output) linear time-invariant stable discrete-time system $\mathbb{G}$ represented by $y_k = \mathbb{G}u_k + d_k, k = 1,2,\cdots$,where u_k is the input and d_k is the system noise. For simplicity, $\mathbb{G}$ is an FIR(finite impulse response) system. In this case,

$$y_k = a_1 u_k + \cdots + a_n u_{k-n+1} + d_k = \boldsymbol{\phi}_k^{\mathrm{T}}\boldsymbol{\theta} + d_k \tag{2.6}$$

where $\boldsymbol{\phi}_k^{\mathrm{T}} = [u_k,\cdots,u_{k-n+1}]$ is the regressor and $\boldsymbol{\theta} = [a_1,\cdots,a_n]^{\mathrm{T}}$ is the parameter vector to be identified and z^{T} denotes the transpose.

The system structure is shown in Figure 2.1, in which the input is finitely quantized with r possible values, $u_k \in \boldsymbol{\mathcal{U}} = \{\mu_1,\cdots,\mu_r\}$. The output y_k is measured by a sensor of m thresholds $-\infty < C_1 < \cdots < C_m < \infty$. The output sensor can be represented by a set of m indicator functions $\boldsymbol{s}_k = [s_k^1,\cdots,s_k^m]^{\mathrm{T}}$,where $s_k^i = I_{\{-\infty < y_k \leq C_i\}}$ for $i = 1,\cdots,m$.

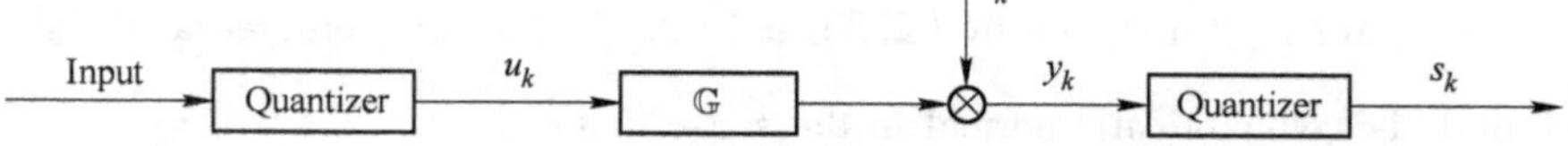

Fig. 2.1 System configuration

Suppose that $\boldsymbol{u} = \{u_k, k = 1,2,\cdots\}$ is an arbitrary input sequence taking quantized values in $\boldsymbol{\mathcal{U}} = \{\mu_1,\cdots,\mu_r\}$. The input u generates a regressor sequence $\{\phi_k^{\mathrm{T}}\}$ that takes values in $l = r^n$ possible (row vector) patterns denoted by $\mathcal{P} = \{\pi_1,\cdots,\pi_l\}$. For example, $\pi_1 = [\mu_1,\cdots,\mu_1,\mu_1]$, $\pi_2 = [\mu_1,\cdots,\mu_1,\mu_2]$,etc. For $k = n+1$, $\cdots, n+N$,we partition the regressor set $\{\phi_{n+1}^{\mathrm{T}},\cdots,\phi_{n+N}^{\mathrm{T}}\}$ according to their patterns π_j. Assume that $\{\phi_{n+1}^{\mathrm{T}},\cdots,\phi_{n+N}^{\mathrm{T}}\}$ contains N_j of pattern π_j , and note that N_j may be zero and

$$\sum_{j=1}^{l} N_j = N$$

The input pattern set is $\{\pi_j : N_j \neq 0\}$,i.e. ,it is the collection of all π_j's that have appeared in $\{\phi_{n+1}^{\mathrm{T}},\cdots,\phi_{n+N}^{\mathrm{T}}\}$.

Assumption 2.2 There exists $\beta_j \geqslant 0$ such that $\lim_{N\to\infty} N_j/N = \beta_j$ for $j \in \boldsymbol{L} = \{1, \cdots, l\}$.

Definition 2.1 We use the following notion.

(1) The pattern π_j is said to be *persistent* if $\beta_j>0$. Without loss of generality, suppose that $\beta_j \neq 0$ for $j \in \boldsymbol{L}_0 = \{1, \cdots, l_0\}$ and $\beta_j = 0$ for $j \in \boldsymbol{L}_0^- = \{l_0+1, \cdots, l\}$.

(2) $\boldsymbol{P}_u = \{\pi_1, \cdots, \pi_{l_0}\}$ is called the *persistent pattern set* of u.

(3) The persistent set P_u of u is said to be *full rank* if the matrix

$$\boldsymbol{\Psi} = \begin{bmatrix} \pi_1 \\ \vdots \\ \pi_{l_0} \end{bmatrix} \in \mathbf{R}^{l_0 \times n}$$

is full column rank.

(4) The input u is said to be *persistently exciting* if P_u is full rank.

Remark 2.1 Non-persistent patterns may exist or appear even infinitely many times. But they do not have impact on asymptotic behavior of the estimates.

Based on the QCCE and weighted least-squares optimization, the identification algorithm is:

(1) At N, if $\pi_j \in \{\pi_j : N_j \neq 0\}$, the observation equations under π_j are

$$y_k^j = w^j + d_k^j, \quad k = 1, \cdots, N_j \tag{2.7}$$

where $w^j = \pi_j \theta$ is unknown and must be estimated. Let

$$\boldsymbol{\nu}^j = [\nu_1^j, \cdots, \nu_m^j]^{\mathrm{T}} \in \mathbf{R}^m \text{ such that } \nu_1^j + \cdots + \nu_m^j = 1 \tag{2.8}$$

and the corresponding QCCE estimate of w^j be denoted by $\hat{w}_{N_j}^j$ with estimation error e_{N_j}, which gives

$$\hat{w}_{N_j}^j = w^j + e_{N_j} \tag{2.9}$$

(2) The weighted vector-valued estimate $\hat{\boldsymbol{W}}_N$ is defined as

$$\hat{\boldsymbol{W}}_N = \begin{bmatrix} \sqrt{N_1}\, w^1 \\ \vdots \\ \sqrt{N_l}\, w^l \end{bmatrix} + \begin{bmatrix} \sqrt{N_1}\, e_{N_1} \\ \vdots \\ \sqrt{N_l}\, e_{N_l} \end{bmatrix} = \boldsymbol{\Phi}_N \theta + \boldsymbol{D}_N$$

where

$$\boldsymbol{\Phi}_N = \begin{bmatrix} \sqrt{N_1}\,\pi_1 \\ \vdots \\ \sqrt{N_l}\,\pi_l \end{bmatrix} \quad \text{and} \quad \boldsymbol{D}_N = \begin{bmatrix} \sqrt{N_1}\,e_{N_1} \\ \vdots \\ \sqrt{N_l}\,e_{N_l} \end{bmatrix}$$

are the weighted regression matrix and the scaled estimation error vector, respectively.

Suppose that $\Phi_N^{\mathrm{T}}\Lambda\Phi_N$ is full rank, where $\Lambda = \mathrm{diag}[\lambda_1, \cdots, \lambda_l] > 0$ will be designed later for improving asymptotic properties and convergence rates. Then, the estimate of θ is

$$\hat{\theta}_N = \left(\frac{1}{N}\Phi_N^{\mathrm{T}}\Lambda\Phi_N\right)^{-1}\frac{1}{N}\Phi_N^{\mathrm{T}}\Lambda\hat{W}_N \tag{2.10}$$

Let $\Sigma(N)$ denote the covariance matrix of the estimation error, i. e., $\Sigma(N) = E(\hat{\theta}_N - \theta)(\hat{\theta}_N - \theta)^{\mathrm{T}}$.

Theorem 2.5 For system (2.6) with quantized observations, if Assumption 2.1 holds and the input u is persistently exciting, then $\hat{\theta}_N$ from (2.10) converges strongly to the true value, i. e., $\hat{\theta}_N \to \theta$ w. p. 1 as $N \to \infty$.

Corollary 2.1 Under the conditions of Theorem 2.5, $\hat{\theta}_N$ is an asymptotically unbiased estimator of θ, i. e., $E\,\hat{\theta}_N \to \theta$ as $N \to \infty$.

Theorem 2.6 Under the conditions of Theorem 2.5, the algorithm (2.10) has the convergence rate

$$\hat{\theta}_N - \theta = O\left(\sqrt{\frac{\log\log N}{N}}\right) \quad \text{w. p. 1} \quad \text{as} \quad N \to \infty$$

Theorem 2.7 Under the conditions of Theorem 2.5, the algorithm (2.10) has the mean-square convergence rate

$$N\Sigma(N) \to (\boldsymbol{\Psi}^{\mathrm{T}}G_1\boldsymbol{\Psi})^{-1}\boldsymbol{\Psi}^{\mathrm{T}}G_2\boldsymbol{\Psi}(\boldsymbol{\Psi}^{\mathrm{T}}G_1\boldsymbol{\Psi})^{-1} \quad \text{as} \quad N \to \infty$$

where $G_1 = \mathrm{diag}[\lambda_1\beta_1, \cdots, \lambda_{l_0}\beta_{l_0}]$, $G_2 = [\beta_1\lambda_1^2\bar{\rho}_1^2, \cdots, \beta_{l_0}\lambda_{l_0}^2\bar{\rho}_{l_0}^2]$, λ_j is the j th diagonal element of Λ and β_j is given by Assumption 2.2 for $j \in L_0$, and $\bar{\rho}_j^2 = (\nu^j)^{\mathrm{T}}$

$U(\pi_j\theta)P(\pi_j\theta)U(\pi_j\theta)\nu^j$ with $U(\cdot)$, $P(\cdot)$ and ν^j being given by (2.3), (2.4) and (2.8).

Theorem 2.8 Under the conditions of Theorem 2.5, the centered and scaled sequence of $\hat{\theta}_N$ is asymptotically normal, i.e.,

$$\sqrt{N}(\hat{\theta}_N-\theta)\xrightarrow{d}\mathcal{N}(0,(\Psi^{\mathrm{T}}G_1\Psi)^{-1}\Psi^{\mathrm{T}}G_2\Psi(\Psi^{\mathrm{T}}G_1\Psi)^{-1})$$

as $N\to\infty$.

Theorem 2.9 The CRLB for estimating θ based on observations of $\{s_k, 1\leqslant k\leqslant N\}$ is

$$\Sigma_{CR}(N)=\left(\sum_{j=1}^{l}N_j\pi_j^{\mathrm{T}}\pi_j\sum_{i=1}^{m+1}\frac{\bar{\zeta}_{i,j}^2}{\zeta_{i,j}}\right)^{-1}$$

where $\zeta_{i,j}(\theta)=F_i(\pi_j\theta)-F_{i-1}(\pi_j\theta)$ and $\bar{\zeta}_{i,j}(\theta)=\partial\zeta_{i,j}/\partial(\pi_j\theta)$, $i=1,\cdots,m+1$, $j\in L$.

Theorem 2.10 Under the conditions of Theorem 2.5, if the optimal QCCE is used in (2.7) ~ (2.9) and

$$\Lambda=\Lambda^*=\mathrm{diag}\left[\sum_{i=1}^{m+1}\frac{\bar{\zeta}_{i,1}^2}{\zeta_{i,1}},\ \cdots,\ \sum_{i=1}^{m+1}\frac{\bar{\zeta}_{i,l_0}^2}{\zeta_{i,l_0}},\ \lambda_{l_0+1},\ \cdots,\ \lambda_l\right]$$

then $\hat{\theta}_N$ from (2.10) is asymptotically efficient in the sense that $N\Sigma(N)-N\Sigma_{CR}(N)\to 0$ as $N\to\infty$.

2.5 Notes

The main content of Section 2.3 comes from [17,39,40], and the one of Section 2.4 does from [41]. The interested reader is referred to them for detailed information.

The empirical-measure-based identification method is only reviewed for a gain system in Section 2.3, and it still can work for FIR systems, rational systems, Wiener systems, Hammerstein systems, systems with Markovian parameters and so on [17,42,43,44,45]. The quantized input-output identification technique in Section 2.4 can also be extended to Wiener systems, Hammerstein systems, and systems with structural uncertainties [46,47,48].

3 FIR System Identification with Scheduled Binary-Valued Observations

This chapter presents a scheduled communication framework for system identification with binary-valued observations. "Sensor scheduling" means the sensor is scheduled to take observation or conduct communication based on some a priori information about the system. The scheduled communication scheme is one of effective ways to deal with the transmission constrain. More specifically, the sensor only transmits its sample data to the remote estimator only when a scheduling condition is satisfied, otherwise, the sensor remains silence. This kind of schemes reduces the communication obviously, meanwhile, it has a better performance than just dropping out packets by a deterministic rule or randomly. Because the not-trigged condition can be known when there is no communication, in this way, the remote estimator can get extra information naturally.

For reducing communication burden while guaranteeing the system performance, taking both quantization and transmission scheduling into account is an intuitive way. However, the scheduling policy breaks the completeness of the observation data, and the quantization makes the relationships between the measured quantized signals and the output to be essentially nonlinear. All these characteristics bring difficulties to the design of parameter estimation algorithms and the analysis of convergence performances.

This chapter focuses on FIR systems to study the identification with scheduled binary-valued observations. Firstly, the formulation of FIR systems is given with binary-valued output observations and the per-specified scheduling policy. By utilizing the statistical property of the system noise, we introduce an empirical-measure-based identification algorithm. Secondly, by use of Bayes' law, total probability formula and the law of large numbers, under periodical input the strong convergence is proved, the mean-square convergence rate and the CRLB are established, and the asymptotic efficiency is also illustrated. Finally, the communication rate is derived and the input design is studied for the minimum communication rate.

The coming sections of this chapter are arranged as follows. Section 3. 1 describes the system set-up and the identification problem with scheduled binary-valued observations. Section 3.2 proposes an algorithm to identify the system parameters. Section 3.3 establishes the strong convergence of the estimates and the mean-square convergence rate of the estimation error, together with the asymptotic efficiency and the communication rate. Section 3.4 uses a numerical example to simulate the main theoretical results obtained and show the effectiveness of the algorithm.

3.1 Problem Formulation

Consider an SISO FIR system

$$y_k = a_1 u_k + \cdots + a_n u_{k-n+1} + d_k = \boldsymbol{\phi}_k^{\mathrm{T}} \boldsymbol{\theta} + d_k \tag{3.1}$$

where $\boldsymbol{\phi}_k = [u_k, \cdots, u_{k-n+1}]^{\mathrm{T}}, \boldsymbol{\theta} = [a_1, \cdots, a_n]^{\mathrm{T}} \in \mathbf{R}^n$ is the unknown parameters, and d_k is the system noise. The output y_k cannot be measured exactly, but done by a binary-valued sensor whose threshold is $C \in (-\infty, +\infty)$. We use an indicator function to represent such binary-valued observation as

$$s_k = I_{\{y_k \leqslant C\}} \tag{3.2}$$

As shown in Figure 3.1, s_k is transmitted to a remote estimator through a communication channel/network. A scheduling policy is implemented, that is,

$$\gamma_k = I_{\{|s_k - \tau_k| \geqslant \delta_k\}} = \begin{cases} 1, & \text{if } |s_k - \tau_k| \geqslant \delta_k \\ 0, & \text{otherwise} \end{cases} \tag{3.3}$$

where $\{\tau_k\}$ and $\{\delta_k\}$ are to be designed to balance the system performance and the communication bandwidth utilization. When $\gamma_k = 1$, a transmission is triggered and then s_k is transmitted to the receiver. When $\gamma_k = 0$, the estimator can not receive anything from the channel. As a consequence, the available information for the estimator is $\{\gamma_k s_k, \gamma_k\}$ at time k.

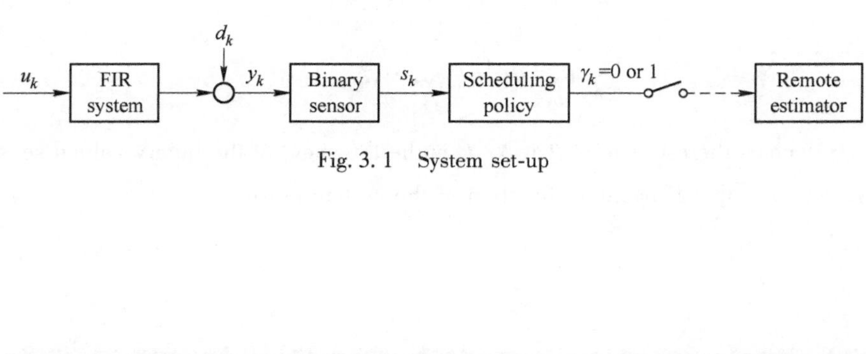

Fig. 3.1 System set-up

This chapter aims to investigate the corresponding identification problem. Two essential issues will be discussed: (1) How to construct algorithms to estimate θ based on the input $\{u_k\}$ and the scheduled binary-valued observation $\{\gamma_k s_k, \gamma_k\}$? (2) How $\{\tau_k\}$ and $\{\delta_k\}$ (namely, the scheduling policy) affect the performance of the algorithm and save the communication resource?

3.2 Identification Algorithm

For simplicity of algorithm design and analysis, it is assumed that the input is n-periodic, that is, $u_{k+n} = u_k$ for all k. This is extensively used in quantized identification and the advantages can be seen in [17]. For non-periodic signals, it can still work (see [49, pp. 81-82]).

Suppose that one-period of $\{u_k\}$ is $u_1 = v_1, u_2 = v_2, \cdots, u_n = v_n$, and denote $\varpi_1 = \phi_1^{\mathrm{T}}, \cdots, \varpi_n = \phi_n^{\mathrm{T}}$. Then the circulant matrix generated by $v_1, \cdots, v_n$ is

$$\boldsymbol{\Phi} = [\varpi_1^{\mathrm{T}}, \cdots, \varpi_n^{\mathrm{T}}]^{\mathrm{T}} \tag{3.4}$$

The n-periodic input $\{u_k\}$ is said to be *full rank* if the n-dimensional matrix $\boldsymbol{\Phi}$ is invertible.

At N (observation length), define

$$N_i = \sum_{j=0}^{L_N-1} I_{\{|1-\tau_{i+jn}| \geq \delta_{i+jn}\}} = \sum_{\substack{j=0,\cdots,L_N-1 \\ |1-\tau_{i+jn}| \geq \delta_{i+jn}}} 1, \quad i = 1, \cdots, n \tag{3.5}$$

where $L_N = \frac{N}{n}$ represents the largest integer less than or equal to $\frac{N}{n}$. We introduce an algorithm to estimate θ as follows

$$\xi_{i,N} = \frac{1}{N_i}\sum_{j=0}^{L_N-1} \gamma_{i+jn} s_{i+jn} \tag{3.6}$$

$$\bar{\xi}_{i,N} = C - F^{-1}(\xi_{i,N}) \tag{3.7}$$

$$\hat{\theta}_N = \boldsymbol{\Phi}^{-1}(\bar{\xi}_{1,N}, \cdots, \bar{\xi}_{n,N})^{\mathrm{T}} \tag{3.8}$$

where $\hat{\theta}_N$ denotes the estimate of θ at N. C is the threshold of the binary-valued sensor in (3.2). $F(\cdot)$ is the distribution function of the system noise.

3.3 Convergence Performance

This section will establish the strong convergence of the algorithm. The mean-square convergence rate of the estimation error will be given, together with the CRLB. Then it will be shown that the algorithm is asymptotically efficient. Moreover, the communication rate will be obtained.

3.3.1 Convergence and Convergence Rate

Theorem 3.1 Consider system (3.1) with binary-valued observations (3.2) and scheduling mechanism (3.3). If the system noise follows Assumption 2.1, the input $\{u_k\}$ is full rank, and $\min_{1\leqslant i\leqslant n}N_i\to\infty$ as $N\to\infty$ (N_i is the one in (3.5)), then the parameter estimate $\hat{\theta}_N$ provided by algorithm (3.6) ~ (3.8) strongly converges to the real value θ, i.e.,

$$\hat{\theta}_N \to \theta,\ \text{w. p. 1.}\quad \text{as}\quad N\to\infty$$

Proof: If $r_k=0$, then we have $\gamma_k s_k=0$. If $r_k=1$ and $|1-\tau_k|<\delta_k$, then it indicates that $s_k=0$ by (3.3), and we also have $\gamma_k s_k=0$. Therefore, it is known that

$$\gamma_k s_k = 0 \quad \text{if} \quad |1-\tau_k| < \delta_k \tag{3.9}$$

According to Bayes' Law (see Item 3 of Appendix A.1.1) and (3.3), one can have

$$\Pr(\gamma_k = 1, s_k = 1) = \Pr(s_k = 1)\Pr(|1-\tau_k| \geqslant \delta_k \mid s_k = 1)$$

Considering that $\Pr(s_k = 1) = \Pr(y_k \leqslant C) = F(C - \phi_k^{\mathrm{T}}\theta)$, it can be seen that

$$E\gamma_k s_k = \Pr(\gamma_k = 1, s_k = 1) = F(C - \phi_k^{\mathrm{T}}\theta) \quad \text{if} \quad |1-\tau_k| \geqslant \delta_k \tag{3.10}$$

Noticing that $\phi_{i+jn}^{\mathrm{T}}=\varpi_i$ for any positive integer j, which and (3.10) can give

$$E\gamma_{i+jn}s_{i+jn} = F(C - \varpi_i\theta) \quad \text{if} \quad |1-\tau_{i+jn}| \geqslant \delta_{i+jn},\ i=1,\cdots,n \tag{3.11}$$

Due to (3.6) and (3.9), it can be derived that

$$\begin{aligned}&\xi_{i,N}\\&=\frac{1}{N_i}\sum_{\substack{j=0,\cdots,L_N-1\\|1-\tau_{i+jn}|\geqslant\delta_{i+jn}}}\gamma_{i+jn}s_{i+jn}+\frac{1}{N_i}\sum_{\substack{j=0,\cdots,L_N-1\\|1-\tau_{i+jn}|<\delta_{i+jn}}}\gamma_{i+jn}s_{i+jn}\\&=\frac{1}{N_i}\sum_{\substack{j=0,\cdots,L_N-1\\|1-\tau_{i+jn}|\geqslant\delta_{i+jn}}}\gamma_{i+jn}s_{i+jn}\end{aligned}\tag{3.12}$$

Since $\min_{1\leqslant i\leqslant n} N_i\to\infty$ if and only if $N_i\to\infty$ for $i=1,\cdots,n$, under hypothesis and by the law of large numbers (see Item 12 of Appendix A. 1. 1 with $\mu=1$), (3. 11) and (3. 12) can yield that $\xi_{i,N}\to F(C-\varpi_i\theta)$, which implies that

$$\bar{\xi}_{i,N}\to\varpi_i\theta,\ \text{w. p. 1}\quad\text{as}\quad N\to\infty,i=1,\cdots,n\tag{3.13}$$

By(3. 7), (3. 8), (3. 13) and (3. 4), we can know that

$$\hat{\theta}_N\to\Phi^{-1}\begin{pmatrix}\varpi_1\\\vdots\\\varpi_n\end{pmatrix}\theta=\theta,\ \text{w. p. 1.}\quad\text{as}\quad N\to\infty$$

Hence, the theorem is proved.

Let $\Sigma(N;\theta)$ represents the covariance matrix of the estimation error of $\hat{\theta}_N$, i. e. ,

$$\Sigma(N;\theta)=E(\hat{\theta}_N-\theta)(\hat{\theta}_N-\theta)^{\mathrm{T}}$$

and denote $\rho_i=\varrho(C-\varpi_i\theta)$ for $i=1,\cdots,n$, where $\varrho(\cdot)$ is given by (2. 2).

Theorem 3. 2 Under the condition of Theorem 3. 1, if

$$\frac{N_i}{N}\to\lambda_i\quad\text{as}\quad N\to\infty,\quad i=1,\cdots,n\tag{3.14}$$

then the mean-square convergence rate of the estimate $\hat{\theta}_N$ given by (3. 8) is

$$N\Sigma(N;\theta)\to\Phi^{-1}\operatorname{diag}\left(\frac{\rho_1}{\lambda_1},\cdots,\frac{\rho_n}{\lambda_n}\right)\Phi^{-T}\quad\text{as}\quad N\to\infty\tag{3.15}$$

Proof: Denoting $\tilde{\xi}_{i,N}=\bar{\xi}_{i,N}-\varpi_i\theta$, from [17] and (3. 13) we can have $N_iE\,\tilde{\xi}_{i,N}^2\to\rho_i$, $\sqrt{N_i}E\,\tilde{\xi}_{i,N}\to0$, $i=1,\cdots,n$, which together with (3. 14) implies that

$$NE[(\tilde{\xi}_{1,N},\cdots,\tilde{\xi}_{n,N})^{\mathrm{T}}(\tilde{\xi}_{1,N},\cdots,\tilde{\xi}_{n,N})]\to\mathrm{diag}\left(\frac{\rho_1}{\lambda_1},\cdots,\frac{\rho_n}{\lambda_n}\right)\text{ as }N\to\infty \quad (3.16)$$

By (3.4), it can be verified that $\theta=\Phi^{-1}(\varpi_1\theta,\cdots,\varpi_n\theta)^{\mathrm{T}}$, which together with (3.8) and (3.16) can lead to

$$\begin{aligned}&N\Sigma(N;\theta)\\&=N\Phi^{-1}E[(\tilde{\xi}_{1,N},\cdots,\tilde{\xi}_{n,N})^{\mathrm{T}}(\tilde{\xi}_{1,N},\cdots,\tilde{\xi}_{n,N})]\,\Phi^{-\mathrm{T}}\\&\to\Phi^{-1}\mathrm{diag}\left(\frac{\rho_1}{\lambda_1},\cdots,\frac{\rho_n}{\lambda_n}\right)\Phi^{-\mathrm{T}}\quad\text{as}\quad N\to\infty\end{aligned}$$

The proof is obtained. □

3.3.2 Asymptotically Efficiency

Lemma 3.1 Based on $\{\gamma_k s_k:1\leqslant k\leqslant N\}$, the CRLB for estimating θ is

$$\sum{}_{CR}(N;\theta)=\left(\Phi^{\mathrm{T}}\mathrm{diag}\left(\frac{N_1}{\rho_1},\cdots,\frac{N_n}{\rho_n}\right)\Phi+\sum_{\substack{k=nL_N+1,\cdots,N\\|1-\tau_k|\geqslant\delta_k}}\frac{\phi_k f^2(C-\phi_k^{\mathrm{T}}\theta)\phi_k^{\mathrm{T}}}{F(C-\phi_k^{\mathrm{T}}\theta)(1-F(C-\phi_k^{\mathrm{T}}\theta))}\right)^{-1} \quad (3.17)$$

Proof: Let z_k be some possible sample value of $\gamma_k s_k$. In view of (3.9), it is known that $z_k=0$ for $k\in\{\iota:1\leqslant\iota\leqslant N,\ |1-\tau_\iota|<\delta_\iota\}$. From this and (3.10), the likelihood function of $\gamma_1 s_1,\cdots,\gamma_N s_N$ taking values $z_1,\cdots,z_N$ conditioned on θ is

$$\begin{aligned}&\ell(z_1,\cdots,z_N;\theta)\\&=\Pr(\gamma_1 s_1=z_1,\cdots,\gamma_N s_N=z_N;\theta)\\&=\prod_{\substack{k=1,\cdots,N\\|1-\tau_k|\geqslant\delta_k}}F^{z_k}(C-\phi_k^{\mathrm{T}}\theta)(1-F(C-\phi_k^{\mathrm{T}}\theta))^{1-z_k}\\&=\prod_{i=1}^{n}\prod_{j\in\mathcal{A}_i}F^{z_{i+jn}}(C-\varpi_i\theta)(1-F(C-\varpi_i\theta))^{1-z_{i+jn}}+\\&\quad\prod_{k\in\mathcal{B}}F^{z_k}(C-\phi_k^{\mathrm{T}}\theta)(1-F(C-\phi_k^{\mathrm{T}}\theta))^{1-z_k}\end{aligned}$$

where $\mathcal{A}_i=\{\iota:0\leqslant\iota\leqslant L_N-1,\ |1-\tau_{i+\iota n}|\geqslant\delta_{i+\iota n}\}$ and $\mathcal{B}=\{\iota:nL_N+1\leqslant\iota\leqslant N,$

$|1-\tau_\iota| \geqslant \delta_\iota\}$

Replace the particular realizations z_k by their corresponding random variables $\gamma_k s_k$, and denote the resulting quantity by $\ell = \ell(\gamma_1 s_1, \cdots, \gamma_N s_N; \theta)$. Let $\mathcal{X}_{i,j} = \gamma_{i+jn} s_{i+jn}$. Then, we have

$$\ell = \prod_{i=1}^{n} \prod_{j \in \mathcal{A}_i} F^{\mathcal{X}_{i,j}}(C - \varpi_i \theta)(1 - F(C - \varpi_i \theta))^{1-\mathcal{X}_{i,j}} +$$

$$\prod_{k \in \mathcal{B}} F^{\gamma_k s_k}(C - \phi_k^{\mathrm{T}} \theta)(1 - F(C - \phi_k^{\mathrm{T}} \theta))^{1-\gamma_k s_k}$$

which results in

$$\log \ell = \sum_{i=1}^{n} \sum_{j \in \mathcal{A}_i} (\mathcal{X}_{i,j} \log F(C - \varpi_i \theta) + (1 - \mathcal{X}_{i,j}) \log(1 - F(C - \varpi_i \theta))) +$$

$$\sum_{k \in \mathcal{B}} (\gamma_k s_k \log F(C - \phi_k^{\mathrm{T}} \theta) + (1 - \gamma_k s_k) \log(1 - F(C - \phi_k^{\mathrm{T}} \theta)))$$

and

$$\frac{\partial \log \ell}{\partial \theta} = \sum_{i=1}^{n} \sum_{j \in \mathcal{A}_i} \left(-\mathcal{X}_{i,j} \varpi_i^{\mathrm{T}} \frac{f(C - \varpi_i \theta)}{F(C - \varpi_i \theta)} + (1 - \mathcal{X}_{i,j}) \varpi_i^{\mathrm{T}} \frac{f(C - \varpi_i \theta)}{1 - F(C - \varpi_i \theta)} \right) +$$

$$\sum_{k \in \mathcal{B}} \left(-\gamma_k s_k \phi_k \frac{f(C - \phi_k^{\mathrm{T}} \theta)}{F(C - \phi_k^{\mathrm{T}} \theta)} + (1 - \gamma_k s_k) \phi_k \frac{f(C - \phi_k^{\mathrm{T}} \theta)}{(1 - F(C - \phi_k^{\mathrm{T}} \theta))} \right)$$

In terms of (3. 10), (3. 5) and (3. 4), it follows that

$$E \frac{\partial^2 \log \ell}{\partial \theta^2} = -\sum_{i=1}^{n} \sum_{j \in \mathcal{A}_i} \frac{\varpi_i^{\mathrm{T}} \varpi_i}{\rho_i} - \sum_{k \in \mathcal{B}} \frac{\phi_k f(C - \phi_k^{\mathrm{T}} \theta) \phi_k^{\mathrm{T}}}{F(C - \phi_k^{\mathrm{T}} \theta)(1 - F(C - \phi_k^{\mathrm{T}} \theta))}$$

$$= -\Phi^{\mathrm{T}} \mathrm{diag}\left(\frac{N_1}{\rho_1}, \cdots, \frac{N_n}{\rho_n} \right) \Phi - \sum_{k \in \mathcal{B}} \frac{\phi_k f^2(C - \phi_k^{\mathrm{T}} \theta) \phi_k^{\mathrm{T}}}{F(C - \phi_k^{\mathrm{T}} \theta)(1 - F(C - \phi_k^{\mathrm{T}} \theta))}$$

Hence, the lemma is proved. □

Theorem 3. 3 Under the condition of Theorem 3. 2, the asymptotic efficiency of the estimate $\hat{\theta}_N$ from (3. 8) can be given by

$$N\Sigma(N;\theta) - N\Sigma_{CR}(N;\theta) \to 0 \quad \text{as} \quad N \to \infty$$

Proof: By virtue of Lemma 3.1, one can have

$$N\sum_{CR}(N;\theta) = (J_{1,N} + J_{2,N})^{-1} \tag{3.18}$$

where

$$J_{1,N} = \frac{1}{N}\Phi^{\mathrm{T}}\mathrm{diag}\left(\frac{N_1}{\rho_1},\cdots,\frac{N_n}{\rho_n}\right)\Phi$$

$$J_{2,N} = \frac{1}{N}\sum_{k\in B}\frac{\phi_k f^2(C-\phi_k^{\mathrm{T}}\theta)\phi_k^{\mathrm{T}}}{F(C-\phi_k^{\mathrm{T}}\theta)(1-F(C-\phi_k^{\mathrm{T}}\theta))}$$

By (3.14), it can be seen that

$$J_{1,N} \to \Phi^{\mathrm{T}}\mathrm{diag}\left(\frac{\lambda_1}{\rho_1},\cdots,\frac{\lambda_n}{\rho_n}\right)\Phi \quad \text{as} \quad N\to\infty \tag{3.19}$$

From Assumption 2.1 we know that $f(\cdot)$ is bounded and then there exists a real number M such that $f(z) \leqslant M$ for $x \in \mathbf{R}$. Note that $\|\phi_k\| \leqslant \tilde{M}$ and $C-\phi_k^{\mathrm{T}}\theta \in [-\overline{M},\overline{M}]$, where $\tilde{M}=n\max_{1\leqslant i\leqslant n}|v_i|$ and $\overline{M}=C+\tilde{M}\|\theta\|^2$ are two constants, and $\|\cdot\|$ is the Euclidean norm of a vector or the Frobenius norm of a matrix. As a result, it can be concluded that

$$\left\|\frac{\phi_k f^2(C-\phi_k^{\mathrm{T}}\theta)\phi_k^{\mathrm{T}}}{F(C-\phi_k^{\mathrm{T}}\theta)(1-F(C-\phi_k^{\mathrm{T}}\theta))}\right\| \leqslant \frac{\tilde{M}^2M^2}{F(-\overline{M})(1-F(\overline{M}))} < \infty$$

Together with $\sum_{k\in B}1 < n$, the above gives rise to

$$\|J_{2,N}\| \leqslant \frac{1}{N}\frac{n\tilde{M}^2M^2}{F(-\overline{M})(1-F(\overline{M}))} \to 0 \quad \text{as} \quad N\to\infty \tag{3.20}$$

Combining (3.18) ~ (3.20), we have

$$N\sum_{CR}(N;\theta)$$

$$\to\left(\Phi^{\mathrm{T}}\mathrm{diag}\left(\frac{\lambda_1}{\rho_1},\cdots,\frac{\lambda_n}{\rho_n}\right)\Phi\right)^{-1}$$

$$=\Phi^{-1}\mathrm{diag}\left(\frac{\rho_1}{\lambda_1},\cdots,\frac{\rho_n}{\lambda_n}\right)\Phi^{-\mathrm{T}} \quad \text{as} \quad N\to\infty$$

which and (3.15) complete the proof. □

3.3.3 Communication Rate

To describe the capacity of the scheduling mechanism in saving the communication resources, we define the communication rate as

$$\bar{\gamma}=\lim_{N\to\infty}\frac{1}{N}\sum_{k=1}^{N}\gamma_k \tag{3.21}$$

Theorem 3.4 Under the condition of Theorem 3.2, if

$$\frac{1}{N}\sum_{j=0}^{L_N-1}I_{\{|\tau_{i+jn}|\geqslant\delta_{i+jn}\}}\to\mu_i,\ i=1,\cdots,n \tag{3.22}$$

then the communication rate $\bar{\gamma}$ from (3.21) can be given by

$$\bar{\gamma}=\sum_{i=1}^{n}(\lambda_i F(C-\varpi_i\theta)+\mu_i(1-F(C-\varpi_i\theta))) \tag{3.23}$$

Proof: By (3.3) and the law of total probability, it can be seen that

$$\begin{aligned}
&E\gamma_{i+jn}\\
&=\Pr(\gamma_{i+jn}=1)\\
&=\Pr(|s_{i+jn}-\tau_{i+jn}|\geqslant\delta_{i+jn})\\
&=\Pr(|s_{i+jn}-\tau_{i+jn}|\geqslant\delta_{i+jn}\mid s_{i+jn}=1)\Pr(s_{i+jn}=1)+\\
&\quad\ \Pr(|s_{i+jn}-\tau_{i+jn}|\geqslant\delta_{i+jn}\mid s_{i+jn}=0)\Pr(s_{i+jn}=0)\\
&=I_{\{|1-\tau_{i+jn}|\geqslant\delta_{i+jn}\}}F(C-\varpi_i\theta)+I_{\{|\tau_{i+jn}|\geqslant\delta_{i+jn}\}}(1-F(C-\varpi_i\theta))
\end{aligned} \tag{3.24}$$

Note that $\lim_{N\to\infty}N/L_N=n$. By (3.5), (3.24) and the law of large numbers, we have

$$\lim_{N\to\infty}\frac{1}{L_N}\sum_{j=0}^{L_N-1}\gamma_{i+jn}$$

$$= \lim_{N\to\infty} \frac{1}{L_N} \sum_{j=0}^{L_N-1} (I_{\{|1-\tau_{i+jn}| \geqslant \delta_{i+jn}\}} F(C - \varpi_i\theta) +$$

$$I_{\{|\tau_{i+jn}| \geqslant \delta_{i+jn}\}} (1 - F(C - \varpi_i\theta)))$$

$$= (1 - F(C - \varpi_i\theta)) \lim_{N\to\infty} \frac{1}{L_N} \sum_{j=0}^{L_N-1} I_{\{|\tau_{i+jn}| \geqslant \delta_{i+jn}\}} +$$

$$F(C - \varpi_i\theta) \lim_{N\to\infty} \frac{N_i}{L_N}$$

$$= n\lambda_i F(C - \varpi_i\theta) + n\mu_i(1 - F(C - \varpi_i\theta))$$

which implies that

$$\lim_{N\to\infty} \frac{1}{N} \sum_{j=0}^{L_N-1} \gamma_{i+jn} = \lim_{N\to\infty} \frac{L_N}{N} \cdot \frac{1}{L_N} \sum_{j=0}^{L_N-1} \gamma_{i+jn}$$

$$= \lambda_i F(C - \varpi_i\theta) + \mu_i(1 - F(C - \varpi_i\theta)) \tag{3.25}$$

Considering $\frac{1}{N}\sum_{i=1}^{N}\gamma_i = \frac{1}{N}\sum_{i=1}^{nL_N}\gamma_i + \frac{1}{N}\sum_{i=nL_N+1}^{N}\gamma_i$ and $\frac{1}{N}\sum_{i=nL_N+1}^{N}\gamma_i \to 0$ as $N\to\infty$, from (3.25) we have

$$\bar{\gamma} = \lim_{N\to\infty} \frac{1}{N}\sum_{i=1}^{nL_N}\gamma_i = \lim_{N\to\infty} \frac{1}{N}\sum_{i=1}^{n}\sum_{j=0}^{L_N-1}\gamma_{i+jn}$$

$$= \sum_{i=1}^{n}\left(\lim_{N\to\infty}\frac{1}{N}\sum_{j=0}^{L_N-1}\gamma_{i+jn}\right)$$

$$= \sum_{i=1}^{n}(\lambda_i F(C - \varpi_i\theta) + \mu_i(1 - F(C - \varpi_i\theta)))$$

The theorem is obtained. □

For a given scheduling policy, by (3.23) we know that $\bar{\gamma}$ is a function of the system input, which can be represented by $\bar{\gamma} = \bar{\gamma}(\varpi_1, \cdots, \varpi_n)$. Then an interesting problem is how to design the input such that $\bar{\gamma}(\varpi_1, \cdots, \varpi_n)$ achieves a minimum value, which can be stated as a constrained minimization problem

$$\min_{\varpi_1,\cdots,\varpi_n}\bar{\gamma}(\varpi_1,\cdots,\varpi_n)$$
$$\text{s. t. } \Phi \text{ is full rank}$$

By (3.23) again, it can be derived that

$$\bar{\gamma}=\sum_{i=1}^{n}((\lambda_i-\mu_i)F(C-\varpi_i\theta)+\mu_i) \tag{3.26}$$

If $\lambda_i \geqslant \mu_i$, then we can have $\min_{\varpi_i}\{(\lambda_i-\mu_i)F(C-\varpi_i\theta)+\mu_i\}=\mu_i$. If $\lambda_i<\mu_i$, then it follows that $\min_{\varpi_i}\{(\lambda_i-\mu_i)F(C-\varpi_i\theta)+\mu_i\}=\lambda_i$. As a consequence, by (3.26) it is known that

$$\min_{\varpi_1,\cdots,\varpi_n}\bar{\gamma}\geqslant\sum_{\substack{i=1,\cdots,n\\ \lambda_i\geqslant\mu_i}}\mu_i+\sum_{\substack{i=1,\cdots,n\\ \lambda_i<\mu_i}}\lambda_i$$

The above provides a lower bound of the minimum communication rate that can be achieved by the input design.

Remark 3.1 In light of (3.5), (3.14) and (3.22), one can see that N_i, λ_i and μ_i are all generated by the scheduled sequences $\{\tau_k\}$ and $\{\delta_k\}$. Therefore, $\min_{1\leqslant i\leqslant n}N_i\to\infty$ in Theorem 3.1, (3.14) in Theorem 3.2 and (3.22) in Theorem 3.4, in fact, present the conditions that should be met by the scheduling policy (3.3) to ensure the strong convergence and the asymptotic efficiency of the identification algorithm (3.6) ~ (3.8) and the existence of the communication rate (3.21).

3.4 Numerical Simulation

Consider an FIR system $y_k=a_1u_k+a_2u_{k-1}+d_k=\phi_k^{\mathrm{T}}\theta+d_k$, where the parameter $\theta=[7,3]^{\mathrm{T}}$ is unknown, but within the range of $\Theta=\{(x,y):0\leqslant x\leqslant 18, 0\leqslant y\leqslant 5\}$. $\{d_k\}$ is i.i.d. with zero mean and standard deviation $\sigma=80$. The binary-valued observation $s_k=I_{\{y_k\leqslant C\}}$, where the threshold $C=8$. The input signal $\{u_k\}$ is 2-periodic with one period $[v_1,v_2]=[23,11]$. The scheduling transmission mechanism is

$$\gamma_k=I_{\{|s_k-\tau_k|\geqslant\delta_k\}}=\begin{cases}1, & \text{if } |s_k-\tau_k|\geqslant\delta_k\\ 0, & \text{otherwise}\end{cases} \tag{3.27}$$

where $\tau_k=F(C-\phi_k^{\mathrm{T}}\bar{\theta})$ with $\bar{\theta}=[9,2.5]^{\mathrm{T}}$, $\delta_k=0.4m_k+0.3$, and $\{m_k\}$ is a maximum length sequence (the coefficients of its feedback function are 0010101001001 and its initial values are 0010010101011).

For the observation length $N=3000$, the identification algorithm (3.6) ~ (3.8) is employed to generate the estimate $\hat{\theta}_N$ of θ. Figure 3.2 displays a trajectory of $\hat{\theta}_N$, where $\hat{\theta}_N$ can indeed converge to the real value and this is accord with Theorem 3.1. The transmission time is demonstrated in Figure 3.3. It can be seen that only 13 measurements are send to the remote estimator in the interval [500,530], which indicates the ability of (3.27) to reduce the communication consumption. Figure 3.4 shows $\frac{1}{N}\sum_{k=1}^{N}\gamma_k$ and

$$\sum_{i=1}^{n}\left(\frac{N_i}{N}F(C-\varpi_i\theta)+\frac{1}{N}\sum_{j=0}^{L_N-1}I_{\{|\tau_{i+jn}|\geqslant\delta_{i+jn}\}}(1-F(C-\varpi_i\theta))\right)$$

with respect to N, where their limits are the same and this is consistent with (3.23) by (3.5) and (3.22). Figure 3.5 gives the relationship between the covariance matrix of the estimation error and the CRLB, which illustrates the asymptotic efficiency of $\hat{\theta}_N$ and Theorem 3.3.

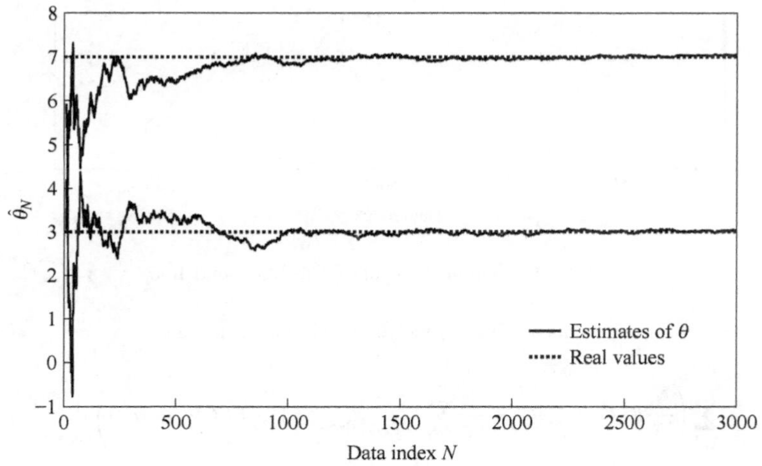

Fig. 3.2 Convergence of $\hat{\theta}_N$

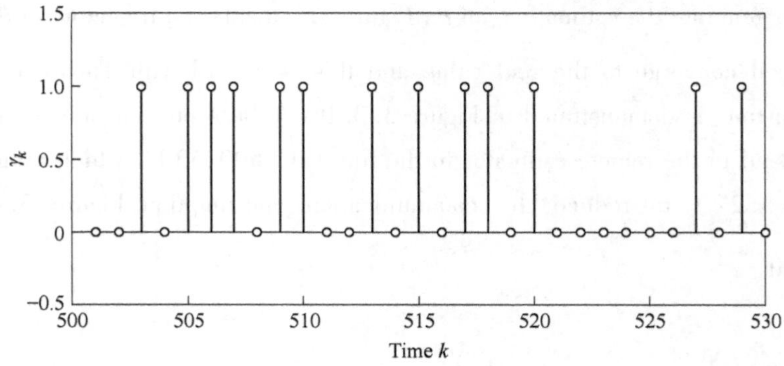

Fig. 3. 3 A realization of γ_k on the horizon [500,530]

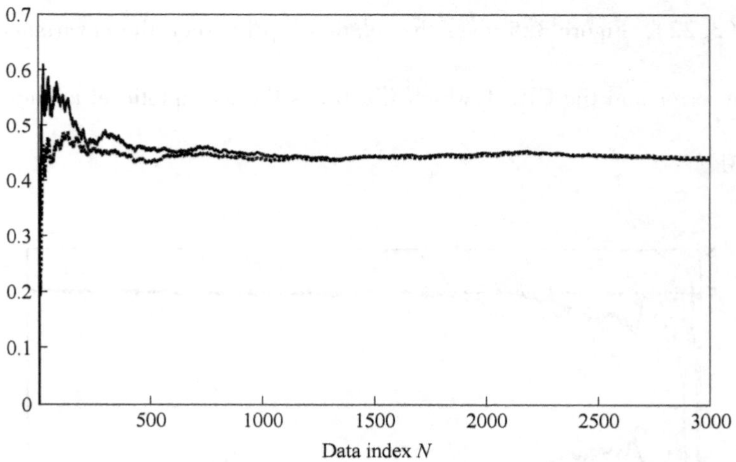

Fig. 3. 4 Communication rate (The blue solid line represents $\frac{1}{N}\sum_{k=1}^{N}\gamma_k$,and the black dotted line does

$$\sum_{i=1}^{n}\left(\frac{N_i}{N}F(C-\varpi_i\theta)+\frac{1}{N}\sum_{j=0}^{L_N-1}I_{\{|\tau_{i+jn}|\geqslant\delta_{i+jn}\}}(1-F(C-\varpi_i\theta))\right)$$)

3. 5 Notes

As the wide applications of networked control systems in the information field, how to save communication resources brings new challenges to the conventional system identifi-

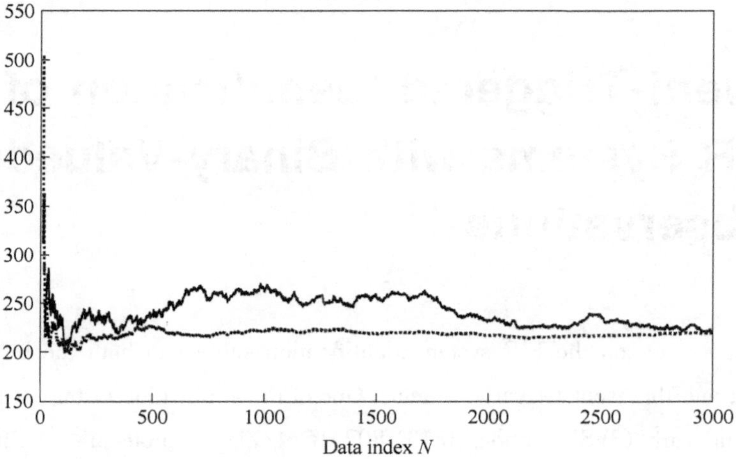

Fig. 3.5 Asymptotic efficiency of $\hat{\theta}_N$: The blue solid line is the average of 200 trajectories of $\| N(\hat{\theta}_N - \theta)(\hat{\theta}_N - \theta)^{\mathrm{T}} \|$ and the black dash one is $\| N\Sigma_{CR}(N;\theta) \|$, where $\| \cdot \|$ is the Frobenius norm

cation. This chapter proposes an identification method for FIR systems with scheduled binary-valued observations. An identification algorithm is provided and proved to be convergent under a periodical input sequence. The convergence rate of the estimation error and the asymptotically efficiency are also obtained. Moreover, the communication rate is discussed.

In this chapter, we discussed a typical way of intermittent communication called scheduled one. It has attracted lots of attention and been extensively introduced to the field of state estimation, feedback control and system identification. For the further study and more details, readers are referred to the references[50~58].

4 Event-Triggered Identification of FIR Systems with Binary-Valued Observations

This chapter focuses on the FIR system identification subject to both the binary-valued quantization and the event-triggered scheme. One of the application examples is the wireless sensor network (WSN) under IEEE 802.15.4/Zigbee protocol[25,59]. In a WSN, there are usually a large amount of sensors transmitting data through wireless channels. Owing to the limited communication and computation ability of the sensor node, the monitored system outputs have to be quantized, and then transmitted to the receiver over a communication channel. To reduce communication burden and improve efficiency, The event-triggered mechanism is often used[61~65]. Another typical example is the networked control system[60,66,67,69~73].

In general, three kinds of methods are often employed for the system identification with quantized data. (1) Empirical-Measure-Based identification method, where the distribution function of the system noise, the output threshold and the quantized data are used to design estimation algorithms. Under periodic inputs/quantized inputs, many excellent results have been obtained[16,17,41]. (2) Maximum Likelihood (ML) estimation. This is based on the joint distribution function of the available information. Due to the quantization nonlinearity, it is always hard to derive an explicit expression of the extremum of the ML function. Together with Expectation Maximization algorithm and Quasi-Newton method, many fundamental progress has been achieved in methodology development, identification algorithms, and convergence properties[68,74~76]. (3) Stochastic Approximation (SA) approach. The key for this method is how to construct an innovation sequence of the available information, choose a suitable step size and develop effective convergence analysis method[77~79].

The event-triggered scheme aims to pick out the "important enough" measurement to be send, and balance the system performance and the communication bandwidth utilization. This entails that the event-triggering condition has to depend on the past information. For example, the difference between the new observation and its conditional mathematical expectation with respect to all the past observations is frequently used to create the event generator, and then the received/available data sequence is strongly correlated. In addition, the quantized output observation cannot provide the exact value, and only can tell us whether or not the output belongs to some given set. The relationships between the measured quantized signals and the input, state and the output are not one-to-one, but essentially nonlinear. Therefore, the coupling of the strong correlation and the high quantization nonlinearity makes that there maybe exist difficulties in maximizing the likelihood function of the event-triggered quantized observations. It also does the algorithm design and the corresponding convergence analysis more difficult and complex than the case of only quantization. In addition, the input in this paper follows a kind of sufficiently rich conditions and does not have periodicity, which restricts the use of the empirical-measure-based method.

The identification algorithm proposed in this chapter belongs to SA type ones. We provide a construction method of the innovation sequence of the event-triggered binary-valued observations based on the a priori information of the unknown parameters and the statistical property of the system noise. By deriving the conditional mathematical expectation of the available data and considering the negative correlations between the innovation and the one-step ahead estimate, it is shown that the estimation error can converge to a finite limit. This together with the sufficient excitations of the inputs yields the strong convergence of the algorithm. Via constructing a martingale difference sequence and calculating its weighted sum, the strong convergence rate of the estimation error is obtained. The event-triggering conditions are also given to ensure the convergence properties of the algorithm. Furthermore, the communication rate is discussed.

The rest of this chapter is organized as follows. In Section 4.1, we describe the formulation of the identification problem. The event-triggered identification algorithm with binary-valued output observations is proposed in Section 4.2. Section 4.3 analyzes the properties of the algorithm, and gives the strong convergence, the convergence rate and the communication rate. In Section 4.4, a numerical example is included to show the effectiveness of the algorithm.

4.1 Problem Formulation

Consider an FIR system:

$$y_k = a_1 u_k + \cdots + a_n u_{k-n+1} + d_k = \boldsymbol{\phi}_k^{\mathrm{T}} \boldsymbol{\theta} + d_k \tag{4.1}$$

where $\boldsymbol{\phi}_k = [u_k, \cdots, u_{k-n+1}]^{\mathrm{T}}, \boldsymbol{\theta} = [a_1, \cdots, a_n]^{\mathrm{T}} \in \mathbf{R}^n$ is the unknown parameters, and d_k is the system noise. The system output y_k is measured by a binary sensor with the known fixed threshold $C \in (-\infty, \infty)$ which can be represented by an indicator function as follows

$$s_k = I_{\{y_k \leqslant C\}} \tag{4.2}$$

As shown in Figure 4.1, a local event detector decides whether the binary-valued output observation s_k is transmitted to the remote estimator. The triggering condition is a kind of send-on-delta mechanism and described by

$$\gamma_k = I_{\{|s_k - \tau_k| \geqslant \delta_k\}} \tag{4.3}$$

where τ_k and δ_k are both $\mathcal{F}_{k-1}$-measurable with $\mathcal{F}_{k-1}$ being the σ -algebra generated by the information received by the estimator before time k, i. e. ,

$$\mathcal{F}_{k-1} = \sigma(\gamma_i, \gamma_i s_i; 1 \leqslant i \leqslant k-1, i \in \mathbf{N}_+)$$

where $\mathbf{N}_+$ is the set of nonnegative integers.

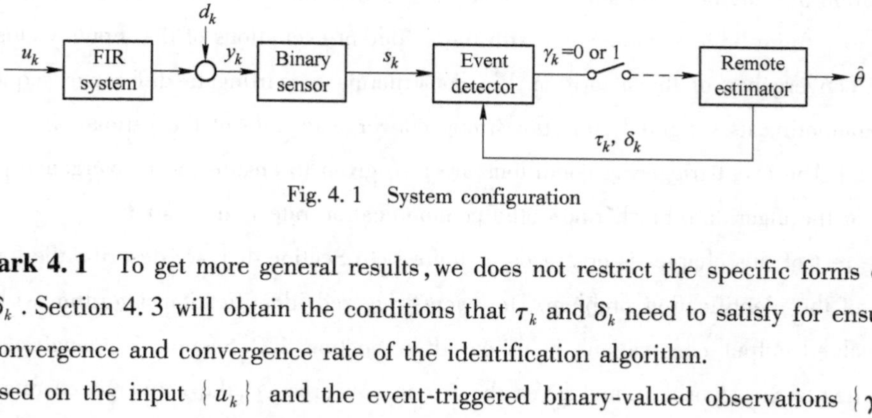

Fig. 4.1 System configuration

Remark 4.1 To get more general results, we does not restrict the specific forms of τ_k and δ_k. Section 4.3 will obtain the conditions that τ_k and δ_k need to satisfy for ensuring the convergence and convergence rate of the identification algorithm.

Based on the input $\{u_k\}$ and the event-triggered binary-valued observations $\{\gamma_k s_k, \gamma_k\}$, this chapter will design algorithms to estimate the unknown parameters θ and discuss how does the binary-valued quantization (4.2) and the event-triggered scheme (4.3) affect the performance of the identification algorithms.

Definition 4.1 Let $\mathcal{M}_N = \{k_i, k_{i+1}, \cdots, k_{i+N-1}\} \subseteq \mathbf{N}_+$ be an N-element subset of $\mathbf{N}_+$ and $\varepsilon > 0$ be a constant. It is said that $\{\phi_k\}$ follows the sufficiently rich condition on $\mathcal{M}_N$ with respect to ε if

$$\sum_{l=0}^{N-1}\phi_{k_{i+l}}\phi_{k_{i+l}}^{\mathrm{T}} \geqslant \varepsilon I_{n\times n}$$

where $I_{n\times n}$ is an $n\times n$ identity matrix. This is denoted by $\{\phi_k\} \in SR(\mathcal{M}_N,\varepsilon)$. Furthermore, if there exist a positive integer N and a constant $\varepsilon>0$ such that

$$\lim_{n\to\infty}\frac{1}{n}\sum_{i=1}^{n} I_{\{\{\phi_k\}\notin SR(\mathcal{M}_{N,i},\varepsilon)\}} = 0$$

where $\{\mathcal{M}_{N,1},\mathcal{M}_{N,2},\cdots\}$ is the set of all N-element subsets of $\mathbf{N}_+$, then it is said that $\{\phi_k\}$ is sufficiently rich in $\mathbf{N}_+$.

Assumption 4.1 The input sequence u is deterministic. The regressor ϕ_k is bounded with

$$\sup_{k\geqslant 1}\|\phi_k\| := M_\phi < \infty \tag{4.4}$$

and there a positive integer N_ϕ and a constant $\varepsilon_\phi>0$ such that and $\{\phi_k\}$ is sufficiently rich in $\mathbf{N}_+$.

Assumption 4.2 There exists a known bounded convex compact set $\Theta\subseteq\mathbf{R}^n$ such that $\theta\in\Theta$ and $\bar{b}=\sup_{z\in\Theta}\|z\|<\infty$, where $\|\cdot\|$ is the Euclidean norm.

4.2 Identification Algorithm

This section will give the recursive identification algorithm. First, the projection operator is defined as follows:

Definition 4.2 For a given convex compact set $\mathbb{S}\subseteq\mathbf{R}^n$, the projection operator $\Pi_{\mathbb{S}}(\cdot)$ is defined as

$$\Pi_{\mathbb{S}}(x) = \arg\min_{w\in\mathbb{S}}\|x-w\|,\ \forall x\in\mathbf{R}^n$$

Lemma 4.1 (Lemma 2.1 in [80]). Given by Definition 4.2, the projection operator $\Pi_{\mathbb{S}}(\cdot)$ is a nonexpansive operator, that is,

$$\|\Pi_{\mathbb{S}}(x)-\Pi_{\mathbb{S}}(y)\| \leqslant \|x-y\|,\ \forall x,y\in\mathbf{R}^n$$

Let $\hat{\theta}_k$ denote the estimate of the unknown parameter θ at time k. A recursive event-triggered identification algorithm is constructed as follows

$$\hat{\theta}_k = \Pi_\Theta\left\{\hat{\theta}_{k-1} + \alpha\frac{\phi_k}{k}\tilde{s}_k\right\} \tag{4.5}$$

$$\tilde{s}_k = I_{\{|1-\tau_k|\geq\delta_k\}}F(C - \phi_k^{\mathrm{T}}\hat{\theta}_{k-1}) - \gamma_k s_k \tag{4.6}$$

where Θ is the domain provided by Assumption 4. 2 and $\Pi_\Theta(\cdot)$ is a projection operator defined by Definition 4. 2 with the convex compact Θ . The initial value $\hat{\theta}_0$ can be arbitrarily chosen in Θ . $F(\cdot)$ is the distribution function of the system noise. C is the threshold in (4. 2), and $\alpha>0$ is a constant scalar.

For convenience, let $\tilde{\theta}_k$ denote the estimation error, i. e. ,

$$\tilde{\theta}_k = \hat{\theta}_k - \theta, \quad k = 0,1,2,\cdots$$

Proposition 4. 1 Under the conditions of Assumptions 4. 1,4. 2 and 2. 1, the following assertions hold.

(1) $E[\tilde{s}_k \mid \mathcal{F}_{k-1}] = I_{\{|1-\tau_k|\geq\delta_k\}}(F(C - \phi_i^{\mathrm{T}}\hat{\theta}_{k-1}) - F(C - \phi_k^{\mathrm{T}}\theta))$.

(2) $\|\tilde{\theta}_i - \tilde{\theta}_j\| \leq \dfrac{(i-j)\alpha M_\phi}{j+1}$ for $i > j \geq 0$.

Proof: (1) From the triggering condition (4. 3), one can get that the received information $r_k s_k$ satisfies

$$\begin{aligned}
&E[r_k s_k \mid \mathcal{F}_{k-1}]\\
&= \Pr\{r_k = 1, s_k = 1 \mid \mathcal{F}_{k-1}\}\\
&= \Pr\{|1-\tau_k| \geq \delta_k, y_k \leq C \mid \mathcal{F}_{k-1}\}\\
&= I_{\{|1-\tau_k|\geq\delta_k\}}\Pr\{d_k \leq C - \phi_k^{\mathrm{T}}\theta \mid \mathcal{F}_{k-1}\}\\
&= I_{\{|1-\tau_k|\geq\delta_k\}}F(C - \phi_k^{\mathrm{T}}\theta)
\end{aligned}$$

In light of the above and (4. 6), it is known that

$$E[\tilde{s}_k \mid \mathcal{F}_{k-1}]$$

$$= E[I_{\{|1-\tau_k| \geq \delta_k\}} F(C - \phi_i^{\mathrm{T}} \hat{\theta}_{k-1}) - \gamma_k s_k \mid \mathcal{F}_{k-1}]$$

$$= I_{\{|1-\tau_k| \geq \delta_k\}} F(C - \phi_i^{\mathrm{T}} \hat{\theta}_{k-1}) - E[\gamma_k s_k \mid \mathcal{F}_{k-1}]$$

$$= I_{\{|1-\tau_k| \geq \delta_k\}} (F(C - \phi_i^{\mathrm{T}} \hat{\theta}_{k-1}) - F(C - \phi_k^{\mathrm{T}} \theta))$$

(2) By (4.5), Assumption 4.1 and $|\tilde{s}_k| \leq 1$, we have

$$\| \tilde{\theta}_i - \tilde{\theta}_j \|$$

$$= \| \hat{\theta}_i - \hat{\theta}_j \|$$

$$= \left\| \sum_{l=j}^{i-1} (\hat{\theta}_{l+1} - \hat{\theta}_l) \right\|$$

$$\leq \sum_{l=j}^{i-1} \| \hat{\theta}_{l+1} - \hat{\theta}_l \| \leq \sum_{l=j}^{i-1} \frac{\alpha}{l+1} \| \phi_{l+1} \tilde{s}_{l+1} \|$$

$$\leq \sum_{l=j}^{i-1} \frac{\alpha M_\phi}{l+1} \leq \frac{(i-j)\alpha M_\phi}{j+1}$$

for $i > j \geq 0$. □

4.3 Properties of the Identification Algorithm

This section will establish the strong convergence and convergence rate of the identification algorithm, and the communication rate will also be analyzed.

4.3.1 Strong Convergence

Definition 4.3 For a subset $\mathcal{M}$ of $\mathbf{N}_+$, if

$$\lim_{n \to \infty} \frac{1}{n} \sum_{i=1}^{n} I_{\{i \in \mathrm{M}\}} = 0$$

then it is said that $\mathcal{M}$ is a zero-density subsequence/set of $\mathbf{N}_+$. Otherwise, it is said that $\mathcal{M}$ is a non-zero-density subsequence/set of $\mathbf{N}_+$.

Lemma 4.2 (Kronecker Lemma, Lemma 2 on Page 114 of [81]). If $\{a_n\}$, $\{b_n\}$ are sequences of real numbers, a_k is increasing to infinity and $\sum_{i=1}^{\infty} b_i/a_i$ converges, then

$$\frac{1}{a_n}\sum_{i=1}^{n} b_i \to 0, \quad n \to \infty$$

Lemma 4.3 (Koopman-Von Neumann Theorem, Theorem 1.2.15 of [82]). If $\{a_k\}$ is a bounded nonnegative sequence, then the sufficient and necessary condition of $\lim_{n\to\infty}\sum_{k=0}^{n-1} a_k = 0$ is the existence of a set of zero density $\mathbb{M}\subset\mathbf{N}_+$ such that $\lim_{\substack{n\to\infty \\ n\notin\mathbb{M}}} a_n = 0$.

Lemma 4.4 Consider two sequences $\{a_k, k\in\mathbf{N}_+\}$ and $\{b_k, k\in\mathbf{N}_+\}$ of non-negative real numbers. If a_k is increasing to infinity,

$$\sum_{k=1}^{\infty}\frac{1}{a_k} = \infty \quad \text{and} \quad \sum_{k=1}^{\infty}\frac{b_k}{a_k} < \infty$$

then there exists a zero-density set $D\subseteq\mathbf{N}_+$ such that

$$\lim_{\substack{k\to\infty \\ k\in\mathcal{D}^-}} b_k = 0$$

where $\mathcal{D}^-$ represents the complementary set of $\mathcal{D}$ in $\mathbf{N}_+$, that is $\mathcal{D}^- = \{k\in\mathbf{N}_+ : k\notin\mathcal{D}\}$.

Proof: Under hypothesis and by virtue of Lemma 4.2, it can be seen that

$$\frac{1}{a_k}\sum_{i=1}^{k} b_i \to 0 \quad \text{as} \quad k\to\infty$$

From the above and Lemma 4.3, the lemma is got. □

Theorem 4.1 Consider system (4.1) with binary-valued observations (4.2) and triggering mechanism (4.3). If Assumptions 4.1 and 4.2 hold, the system noise follows Assumption 2.1 and there exist a non-zero-density subsequence $\{k_i, i\in\mathbf{N}_+\}$ of N_+ and a constant $\iota>0$ such that

$$|1-\tau_{k_i}| \geqslant \delta_{k_i} \quad \text{and} \quad \limsup_{k_i\to\infty}\frac{k_{i+\iota}-k_i}{k_i} = 0 \tag{4.7}$$

then the parameter estimate $\hat{\theta}_k$ provided by the algorithm (4.5) and (4.6) can strongly converge to the real value, i.e.,

$$\lim_{k\to\infty}\hat{\theta}_k = \theta, \text{w. p. } 1, \quad k \to \infty$$

Proof: Note that Θ is a convex compact set by Assumption 4.2. From (4.5) and Lemma 4.1, it can be obtained that

$$\begin{aligned}\|\tilde{\theta}_k\| &= \|\hat{\theta}_k - \theta\| \\ &= \left\| \Pi_\Theta\left(\hat{\theta}_{k-1} + \alpha\frac{\phi_k}{k}\tilde{s}_k\right) - \Pi_\Theta(\theta) \right\| \\ &\leqslant \left\| \left(\hat{\theta}_{k-1} + \alpha\frac{\phi_k}{k}\tilde{s}_k\right) - \theta \right\| \\ &= \left\| \tilde{\theta}_{k-1} + \alpha\frac{\phi_k}{k}\tilde{s}_k \right\|\end{aligned}$$

which together with $\tilde{s}_k^2 \leqslant 1$ implies that

$$\|\tilde{\theta}_k\|^2 \leqslant \|\tilde{\theta}_{k-1}\|^2 + \frac{2\alpha}{k}\phi_k^T\tilde{\theta}_{k-1}\tilde{s}_k + \alpha^2\frac{\|\phi_k\|^2}{k^2} \tag{4.8}$$

Since the input sequence is deterministic, one can get

$$\begin{aligned}&E[\|\tilde{\theta}_k\|^2 \mid \mathcal{F}_{k-1}] \\ &\leqslant E[\|\tilde{\theta}_{k-1}\|^2 \mid \mathcal{F}_{k-1}] + \frac{2\alpha}{k}E[\phi_k^T\tilde{\theta}_{k-1}\tilde{s}_k \mid \mathcal{F}_{k-1}] + \alpha^2\frac{\|\phi_k\|^2}{k^2} \\ &= \|\tilde{\theta}_{k-1}\|^2 + \frac{2\alpha}{k}\phi_k^T\tilde{\theta}_{k-1}E[\tilde{s}_k \mid \mathcal{F}_{k-1}] + \alpha^2\frac{\|\phi_k\|^2}{k^2}\end{aligned} \tag{4.9}$$

According to Differential Mean Value Theorem, we have $F(C - \phi_k^T\hat{\theta}_{k-1}) - F(C - \phi_k^T\theta) = -f(\xi_k)\phi_k^T\tilde{\theta}_{k-1}$, where ξ_k is between $C - \phi_k^T\hat{\theta}_{k-1}$ and $C - \phi_k^T\theta$ and $f(\cdot)$ is the density function of d_k known from Assumption 2.1, i.e., $f(x) = \mathrm{d}F(x)/\mathrm{d}x$. By (1) of Proposition 4.1, it follows that

$$E[\tilde{s}_k \mid \mathcal{F}_{k-1}] = -I_{\{|1-\tau_k| \geqslant \delta_k\}}f(\xi_k)\phi_k^T\tilde{\theta}_{k-1} \tag{4.10}$$

From Assumption 4.2 and (4.4), it is known that $|\xi_k| \leqslant |C| + \bar{b}M_\phi$. By (4.9) and (4.10), we can obtain

$$
\begin{aligned}
&E[\|\tilde{\theta}_k\|^2 \mid \mathcal{F}_{k-1}] \\
&\leqslant \|\tilde{\theta}_{k-1}\|^2 - \frac{2\alpha}{k} I_{\{|1-\tau_k| \geqslant \delta_k\}} f(\xi_k)(\phi_k^{\mathrm{T}}\tilde{\theta}_{k-1})^2 + \alpha^2 \frac{\|\phi_k\|^2}{k^2} \\
&\leqslant \|\tilde{\theta}_{k-1}\|^2 - \frac{2\alpha f(|C| + \bar{b}M_\phi)}{k} I_{\{|1-\tau_k| \geqslant \delta_k\}} (\phi_k^{\mathrm{T}}\tilde{\theta}_{k-1})^2 + \alpha^2 \frac{\|\phi_k\|^2}{k^2}
\end{aligned}
\tag{4.11}
$$

By (2) of Proposition 4.1, one can derive

$$
\begin{aligned}
&E\|\tilde{\theta}_k\|^2 \\
&= E[E[\|\tilde{\theta}_k\|^2 \mid \mathcal{F}_{k-1}]] \\
&\leqslant E\|\tilde{\theta}_{k-1}\|^2 + \alpha^2 \frac{\|\phi_k\|^2}{k^2} - \frac{2\alpha f(|C| + \bar{b}M_\phi)}{k} E(I_{\{|1-\tau_k| \geqslant \delta_k\}} (\phi_k^{\mathrm{T}}\tilde{\theta}_{k-1})^2) \\
&\leqslant E\|\tilde{\theta}_0\|^2 + \alpha^2 \sum_{i=1}^{k} \frac{\|\phi_i\|^2}{i^2} - \sum_{i=1}^{k} \frac{2\alpha f(|C| + \bar{b}M_\phi)}{i} E(I_{\{|1-\tau_i| \geqslant \delta_i\}} (\phi_i^{\mathrm{T}}\tilde{\theta}_{i-1})^2)
\end{aligned}
\tag{4.12}
$$

Together with $0 \leqslant E\|\tilde{\theta}_k\|^2 \leqslant 4\bar{b}^2$ and

$$
\sum_{i=1}^{\infty} \frac{\|\phi_i\|^2}{i^2} \leqslant M_\phi^2 \sum_{i=1}^{\infty} \frac{1}{i^2} = \frac{M_\phi^2 \pi^2}{6} < \infty \tag{4.13}
$$

it is known that (4.12) indicates

$$
\sum_{k=1}^{\infty} E\left(\frac{I_{\{|1-\tau_k| \geqslant \delta_k\}}}{k} (\phi_k^{\mathrm{T}}\tilde{\theta}_{k-1})^2\right) < \infty
$$

and

$$
\sum_{k=1}^{\infty} \frac{I_{\{|1-\tau_k| \geqslant \delta_k\}}}{k} (\phi_k^{\mathrm{T}}\tilde{\theta}_{k-1})^2 < \infty \text{, w. p. 1} \tag{4.14}
$$

Let $a_k = k$ and $b_k = I_{\{|1-\tau_k| \geqslant \delta_k\}}(\phi_k^{\mathrm{T}}\tilde{\theta}_{k-1})^2$ in Lemma 4.4. On account of (4.14), we know that there exists a zero-density set $\mathcal{M} \subseteq \mathbf{N}_+$ such that

$$I_{\{|1-\tau_k| \geqslant \delta_k\}}(\phi_k^{\mathrm{T}}\tilde{\theta}_{k-1})^2 \to 0, \quad k \in \mathcal{M}^-, k \to \infty$$

Noticing that $\{\phi_k\}$ is sufficiently rich in $\mathbf{N}_+$ by Assumption 4.1 and the density of $\{k_i\}$ is non zero, there exist a positive integer $\overline{N}$, a constant $\overline{\varepsilon} > 0$ and a subsequence $\{\nu_i\}$ of $\{k_i\}$ such that

$$|1 - \tau_{\nu_i}| \geqslant \delta_{\nu_i} = 1, \quad \limsup_{\nu_i \to \infty} \frac{\nu_{i+\overline{N}-1} - \nu_i}{\nu_i} \to 0$$

$$\text{and} \quad \{\phi_k\} \in SR(\{\nu_i, \nu_{i+1}, \cdots, \nu_{i+\overline{N}-1}\}, \overline{\varepsilon}) \tag{4.15}$$

for $i = 1, 2, \cdots$. Then we have

$$(\phi_{\nu_i}^{\mathrm{T}}\tilde{\theta}_{\nu_i-1})^2 \to 0, \text{w. p. } 1, \quad i \to \infty \tag{4.16}$$

For $j = i, i+1, \cdots, i+\overline{N}-1$, we have

$$\begin{aligned}
&(\phi_{\nu_j}^{\mathrm{T}}\tilde{\theta}_{\nu_i-1})^2 \\
&= (\phi_{\nu_j}^{\mathrm{T}}(\tilde{\theta}_{\nu_i-1} - \tilde{\theta}_{\nu_j-1} + \tilde{\theta}_{\nu_j-1}))^2 \\
&= (\phi_{\nu_j}^{\mathrm{T}}\tilde{\theta}_{\nu_j-1})^2 + (\phi_{\nu_j}^{\mathrm{T}}(\tilde{\theta}_{\nu_i-1} - \tilde{\theta}_{\nu_j-1}))^2 + 2(\tilde{\theta}_{\nu_j-1}^{\mathrm{T}}\phi_{\nu_j}\phi_{\nu_j}^{\mathrm{T}}(\tilde{\theta}_{\nu_i-1} - \tilde{\theta}_{\nu_j-1}))
\end{aligned} \tag{4.17}$$

By use of (2) in Proposition 4.1, it can see that

$$\begin{aligned}
(\phi_{\nu_j}^{\mathrm{T}}(\tilde{\theta}_{\nu_i-1} - \tilde{\theta}_{\nu_j-1}))^2 &\leqslant \frac{\alpha^2 M_\phi^4 (\nu_j - \nu_i)^2}{\nu_i} \\
&\leqslant \frac{\alpha^2 M_\phi^4 (\nu_{i+\overline{N}-1} - \nu_i)^2}{\nu_i^2}
\end{aligned} \tag{4.18}$$

and

$$\tilde{\theta}^{\mathrm{T}}_{\nu_j-1}\phi_{\nu_j}\phi^{\mathrm{T}}_{\nu_j}(\tilde{\theta}_{\nu_i-1} - \tilde{\theta}_{\nu_j-1}) \leqslant \frac{\alpha\bar{b}M^3_{\phi}(\nu_j - \nu_i)}{\nu_i}$$

$$\leqslant \frac{\alpha\bar{b}M^3_{\phi}(\nu_{i+\bar{N}-1} - \nu_i)}{\nu_i} \tag{4.19}$$

Due to (4.18)、(4.19) and (4.15), one can obtain that $(\phi^{\mathrm{T}}_{\nu_j}(\tilde{\theta}_{\nu_i-1} - \tilde{\theta}_{\nu_j-1}))^2 \to 0$ and $(\tilde{\theta}^{\mathrm{T}}_{\nu_j-1}\phi_{\nu_j}\phi^{\mathrm{T}}_{\nu_j}(\tilde{\theta}_{\nu_i-1} - \tilde{\theta}_{\nu_j-1})) \to 0$ as $i \to \infty$. With (4.16) and (4.17), we have

$$(\phi^{\mathrm{T}}_{\nu_j}\tilde{\theta}_{\nu_i-1})^2 \to 0 \quad \text{for} \quad j = i, i+1, \cdots, i+\bar{N}-1 \tag{4.20}$$

By (4.15), it is known that

$$\bar{\varepsilon}\,\|\tilde{\theta}_{\nu_i-1}\|^2 \leqslant \tilde{\theta}^{\mathrm{T}}_{\nu_i-1}\Big(\sum_{j=i}^{i+\bar{N}-1}\phi_{\nu_j}\phi^{\mathrm{T}}_{\nu_j}\Big)\tilde{\theta}_{\nu_i-1}$$

$$= \sum_{j=i}^{i+\bar{N}-1}(\phi^{\mathrm{T}}_{\nu_j}\tilde{\theta}^{\mathrm{T}}_{\nu_i-1})^2 \tag{4.21}$$

The above and (4.20) can yield $\|\tilde{\theta}_{\nu_i-1}\|^2 \to 0, i \to \infty$.

By (4.11), we obtain $E[\,\|\tilde{\theta}_k\|^2 \mid \mathcal{F}_{k-1}] \leqslant \|\tilde{\theta}_{k-1}\|^2 + \alpha^2\dfrac{\|\phi_k\|^2}{k^2}$, which together with (4.13) and Lemma A.3 implies that $\|\tilde{\theta}_k\|$ converges to a bounded limit a.s. With $\lim\limits_{i\to\infty}\|\tilde{\theta}_{\nu_i-1}\|^2 = 0$, it can be got that $\lim\limits_{k\to\infty}\tilde{\theta}_k = 0$ and hence the theorem is proved. □

4.3.2 Convergence Rate

Theorem 4.2 Under the condition of Theorem 4.1, if (4.7) is constrained to

$$|1 - \tau_{k_i}| \geqslant \delta_{k_i} \text{ and } \bar{K} := \sup_{i\in\mathbf{N}_+}(k_{i+\iota} - k_i) < \infty \tag{4.22}$$

and

$$\alpha > \frac{M^2_{\phi}\bar{K}\left\lceil \dfrac{N_{\phi}}{\iota} \right\rceil}{2f(|C| + \bar{b}M_{\phi})\varepsilon_{\phi}} \tag{4.23}$$

then the algorithm (4.5) and (4.6) has the strong convergence rate

$$\| \tilde{\theta}_k \|^2 = O\left(\frac{\log k}{k}\right)$$

where z denotes the least integer greater than or equal to z.

Proof: From (4.4), (4.6) and (4.8), it can be obtained that

$$k \| \tilde{\theta}_k \|^2 - (k-1) \| \tilde{\theta}_{k-1} \|^2$$

$$\leqslant \| \tilde{\theta}_{k-1} \|^2 + \alpha^2 \frac{M_\phi^2}{k} - 2\alpha f(\xi_k) I_{\{|1-\tau_k| \geqslant \delta_k\}} (\phi_k^{\mathrm{T}} \tilde{\theta}_{k-1})^2 +$$

$$2\alpha \phi_k^{\mathrm{T}} \tilde{\theta}_{k-1} (I_{\{|1-\tau_k| \geqslant \delta_k\}} F(C - \phi_k^{\mathrm{T}} \theta) - \gamma_k s_k)$$

Noticing $|\xi_k| \leqslant |C| + \bar{b} M_\phi$ again, the above can yield

$$k \| \tilde{\theta}_k \|^2 \leqslant \sum_{i=1}^{k} \alpha^2 \frac{M_\phi^2}{i} + J_1(k) - 2\alpha f(|C| + \bar{b} M_\phi) J_2(k) + 2\alpha J_3(k) \tag{4.24}$$

with

$$J_1(k) = \sum_{i=1}^{k} \| \tilde{\theta}_{i-1} \|^2 \tag{4.25}$$

$$J_2(k) = \sum_{i=1}^{k} I_{\{|1-\tau_i| \geqslant \delta_i\}} (\phi_i^{\mathrm{T}} \tilde{\theta}_{i-1})^2 \tag{4.26}$$

$$J_3(k) = \sum_{i=1}^{k} \phi_i^{\mathrm{T}} \tilde{\theta}_{i-1} (I_{\{|1-\tau_i| \geqslant \delta_i\}} F(C - \phi_i^{\mathrm{T}} \theta) - \gamma_i s_i) \tag{4.27}$$

Considering the condition of the theorem, we know that there must exist a non-zero-density subsequence $\{\eta_i\}$ of $\{k_i\}$ such that

$$|1 - \tau_{\eta_i}| \geqslant \delta_{\eta_i} = 1, \quad \sup_{i \in \mathbf{N}_+} (\eta_{i+\max\{N_\phi, \iota\}} - \eta_i)$$
$$\text{and} \quad \{\phi_k\} \in SR(\{\eta_i, \eta_{i+1}, \cdots, \eta_{i+\max\{N_\phi, \iota\}-1}\}, \varepsilon_\phi) \tag{4.28}$$

Denote $\bar{i}_\eta(k) = \arg\max\{i \in \mathbf{N}_+ : \eta_i \leqslant k\}$. Employing the technique in (4.17) ~ (4.19) and (4.21), by (4.25)、(4.26) and (4.28) it can be derived that

$$
\begin{aligned}
J_1(k) &\leqslant \sum_{i=1}^{k_{\bar{i}_\eta(k)}} \|\tilde{\theta}_{i-1}\|^2 + O(1) \\
&\leqslant \sum_{l=1}^{\frac{\bar{i}_\eta(k)}{\max\{N_\phi,\iota\}}} \sum_{j=1+(l-1)\max\{N_\phi,\iota\}}^{l\max\{N_\phi,\iota\}} (\phi_j^{\mathrm{T}}\tilde{\theta}_{j-1})^2 + O(1) \\
&\leqslant \bar{K}\frac{N_\phi}{\iota}\Delta(k) + O(\log k)
\end{aligned} \tag{4.29}
$$

and

$$
\begin{aligned}
J_2(k) &\geqslant \sum_{i=1}^{\bar{i}_\eta(k)} (\phi_{\eta_i}^{\mathrm{T}}\tilde{\theta}_{\eta_i-1})^2 \\
&\geqslant \sum_{l=1}^{\frac{\bar{i}_\eta(k)}{\max\{N_\phi,\iota\}}} \sum_{j=1+(l-1)\max\{N_\phi,\iota\}}^{l\max\{N_\phi,\iota\}} (\phi_{\eta_j}^{\mathrm{T}}\tilde{\theta}_{\eta_j-1})^2 \\
&\geqslant \varepsilon_\phi\Delta(k) + O(\log k)
\end{aligned} \tag{4.30}
$$

with $\Delta(k) = \sum_{l=1}^{\frac{\bar{i}_\eta(k)}{\max\{N_\phi,\iota\}}} \|\tilde{\theta}_{\eta_{1+(l-1)\max\{N_\phi,\iota\}}-1}\|^2$. In light of (2) of Proposition 4.1, it is known that $\{I_{\{|1-\tau_i|\geqslant\delta_i\}} F(C-\phi_i^{\mathrm{T}}\theta) - \gamma_i s_i\}, \mathcal{F}_i, i \geqslant 1\}$ is a martingale difference sequence. According to (4.27) and (1) of Lemma A.4, for any $\zeta > 1/2$ one can have

$$
J_3(k) = O\left(\sqrt{\sum_{i=1}^{k} (\phi_i^{\mathrm{T}}\tilde{\theta}_{i-1})^2}\left(\log\sqrt{\sum_{i=1}^{k} (\phi_i^{\mathrm{T}}\tilde{\theta}_{i-1})^2}\right)^\zeta\right) \tag{4.31}
$$

Since the density of $\{\eta_i\}$ is not zero, (4.24) and (4.29) ~ (4.31) can give

$$
k\|\tilde{\theta}_k\|^2 \leqslant O(\log k) + \left(\bar{K}\frac{N_\phi}{\iota} - 2\alpha f(|C| + \bar{b}M_\phi)\varepsilon_\phi + o(1)\right)\Delta(k)
$$

Under (4.23), the theorem is proved. □

4.3.3 Implementation of the Event-Triggered Mechanism

From Theorem 4.1 and Theorem 4.2, it can be seen that the event generator (4.3) needs to follow the condition (4.7) or (4.22) for guaranteeing the convergence property of the identification algorithm. Then, what τ_k and δ_k can yield (4.7) or (4.22)? Since τ_k and δ_k are $\mathcal{F}_{k-1}$-measurable, we can always design them to get (4.7) or (4.22). In practical use, τ_k can be a predictor (or an estimator) of s_k in some sense, and it is usually the conditional mathematical expectation of s_k with respect to $\mathcal{F}_{k-1}$, i.e., the minimum mean-squared error estimator of s_k based on all the past information. The next provides a way to do so.

For a given $\iota \in \mathbf{N}_+$, choose a non-zero-density subsequence of $\mathbf{N}_+$, denoted by $\mathbb{K} = \{k_i\}$, such that

$$\limsup_{k_i \to \infty} \frac{k_{i+\iota} - k_i}{k_i} = 0 \tag{4.32}$$

or

$$\sup_{i \in \mathbf{N}_+} (k_{i+\iota} - k_i) < \infty \tag{4.33}$$

Considering $E[s_k \mid \mathcal{F}_{k-1}] = F(C - \phi_k^{\mathrm{T}} \hat{\theta}_{k-1})$, we make τ_k and δ_k as

$$\tau_k = F(C - \phi_k^{\mathrm{T}} \hat{\theta}_{k-1}) \tag{4.34}$$

$$\delta_k = \mu_k \left(I_{\{k \notin \mathbb{K}\}} + \mid 1 - F(C - \phi_k^{\mathrm{T}} \hat{\theta}_{k-1}) I_{\{k \in \mathbb{K}\}} \mid \right) \tag{4.35}$$

where $\mu_k \in (0,1)$ is $\mathcal{F}_{k-1}$-measurable.

It can be verified that both τ_k and δ_k are $\mathcal{F}_{k-1}$-measurable, and $|1-\tau_k| \geqslant \delta_k$ for $k \in \mathbb{K}$. And then one can know that (4.7) holds under (4.32) and (4.22) hold sunder (4.33), which indicates that (4.34) and (4.35) actually gives an implementation of the event-triggered mechanism (4.3) with (4.7) or (4.22).

Remark 4.2 In (4.34) and (4.35), the current estimate can be sent from data fusion center to sensors by downlink channel. In most of network environments, the abilities of upstream and downstream are different. Such as in wireless sensor network, transmitting radio frequency signals is the most power-consuming part for a sensor. On the contrary, receiving signals is much more energy saving. Thus, reducing times and data size of transmission is important for extending life span of power limited wireless sensors. But the power of data fusion center is not limited, hence reducing transmission through the event-triggered mechanism by the use of feedback information is cost-efficient. The IEEE 802.15.4/Zigbee protocol is one example, where the coordinator can broadcasts the required feedback information to sensors at the beginning of each periodic frame. Such mechanism can also be found in literatures on the state estimation[83~85] and etc.

4.3.4 Communication Rate

Theorem 4.3 If Assumptions 4.1, 4.2 and 2.1 hold and the communication rate $\overline{\gamma}$ from (4.3) exists, then one can have

$$\overline{\gamma} = \lim_{k\to\infty}\frac{1}{k}\sum_{i=1}^{k}\{I_{\{|1-\tau_i|\geqslant\delta_i\}}F(C-\phi_i^{\mathrm{T}}\theta)+I_{\{|\tau_i|\geqslant\delta_i\}}(1-F(C-\phi_i^{\mathrm{T}}\theta))\},\text{w. p. 1}$$

where $\overline{\gamma}$ is defined by (1.2)

Proof: Since τ_k and δ_k are $\mathcal{F}_{k-1}$-measurable, we have

$$\begin{aligned}
&E[\gamma_k \mid \mathcal{F}_{k-1}]\\
&=\Pr(\gamma_k=1\mid\mathcal{F}_{k-1})\\
&=\Pr(|s_k-\tau_k|\geqslant\delta_k\mid\mathcal{F}_{k-1})\\
&=I_{\{|1-\tau_k|\geqslant\delta_k\}}\Pr(s_k=1\mid\mathcal{F}_{k-1})+I_{\{|\tau_k|\geqslant\delta_k\}}\Pr(s_k=0\mid\mathcal{F}_{k-1})\\
&=I_{\{|1-\tau_k|\geqslant\delta_k\}}F(C-\phi_k^{\mathrm{T}}\theta)+I_{\{|\tau_k|\geqslant\delta_k\}}(1-F(C-\phi_k^{\mathrm{T}}\theta))
\end{aligned}$$

which illustrates that $\{\tilde{\gamma}_k,\mathcal{F}_k\}$ is a martingale difference sequence, where

$$\tilde{\gamma}_k:=\gamma_k-I_{\{|1-\tau_k|\geqslant\delta_k\}}F(C-\phi_k^{\mathrm{T}}\theta)-I_{\{|\tau_k|\geqslant\delta_k\}}(1-F(C-\phi_k^{\mathrm{T}}\theta))$$

Consequently, it can verified that

$$E\Big(\sum_{i=1}^{k}\tilde{\gamma}_i\Big)^2\leqslant 9k^2<\infty \quad\text{and}\quad \sum_{i=1}^{\infty}\frac{E(\tilde{\gamma}_{i-1})^2}{i^2}<\infty$$

By virtue of Lemma A.5, one can get $\tilde{\gamma}_k/k \to 0$, w. p. 1, $k\to\infty$ and hence the proof of the theorem is completed. □

4.4 Numerical Simulation

We consider an FIR system $y_k = \phi_k^T\theta + d_k$ with the binary-valued output observations $s_k = I_{\{y_k \leqslant C\}}$, where the unknown parameter vector $\theta = [5, 15]^T$ known as in $\Theta = \{(x, y): |x| \leqslant 30, |y| \leqslant 30\}$, the threshold $C = 8$, and $\{d_k\}$ is a sequence of i. i. d. normal random variables with zero mean and standard deviation $\sigma = 6$. The regressor $\phi_k = [\sin(k), \sin(k-1)]^T$ and follows Assumption 4.1 with $N_\phi = 2$ and $M_\phi \leqslant \sqrt{2}$. The event-triggered transmission mechanism is $\gamma_k = I_{\{|s_k - \tau_k| \geqslant \delta_k\}}$ where $\{\tau_k\}$ and $\{\delta_k\}$ are given by (4.34) and (4.35) with $\mu_k = 0.4$. For choosing $\mathbb{K} = \{k_i\}$, a 13-chip maximum length sequence is generated (the coefficients of its feedback function are 0010101001001 and its initial values are 0010010101011) and the time of the fifth '1' from every five '1's is selected as the element of $\mathbb{K}$. Then the density of $\mathbb{K}$ is non-zero and it can be calculated that $\iota = 1$, $\overline{K} = 23$ and $\varepsilon_\phi = 0.0046$.

The identification algorithm (4.5) and (4.6) with $\alpha = 170$ is employed to make the estimate $\hat{\theta}_k$ of θ. The transmission time and the average communication rate are illustrated in Figure 4.2 and Figure 4.3. It can be seen that only 18 measurements are send to the estimator in the interval [300, 400], and $\overline{\gamma}$ is close to 0.2, which indicated the better ability to reduce the communication consumption.

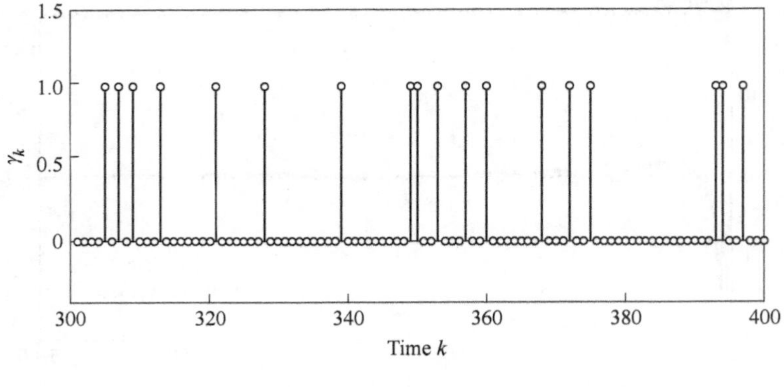

Fig. 4.2 A realization of γ_k on the horizon [300, 400]

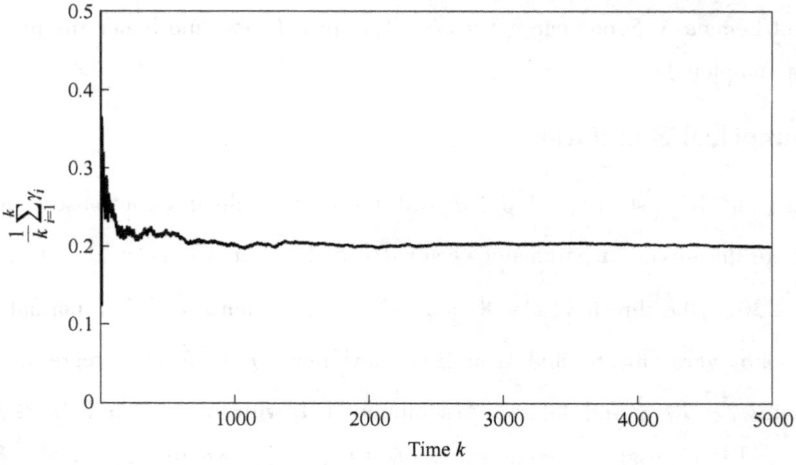

Fig. 4. 3 Communication rate

Figure 4. 4 shows a trajectory of $\hat{\theta}_k$. One can see that $\hat{\theta}_k$ can indeed converge to the real value. Figure 4. 5 demonstrates that $k \| \tilde{\theta}_k \|^2/\log k$ is bounded, which implies that the convergence rate of $\| \tilde{\theta}_k \|^2$ has the order of $\log k/k$. Moreover, Figure 4. 5 also displays the comparison between (4. 5) and (4. 6) and the algorithm without the event-triggered transmission mechanism (i. e. , $\gamma_k \equiv 1$). The curve of the former is higher than the one of the latter. As we can imagine, although the event-triggered scheme reduces the communication rate, it also causes influences on the convergence rate.

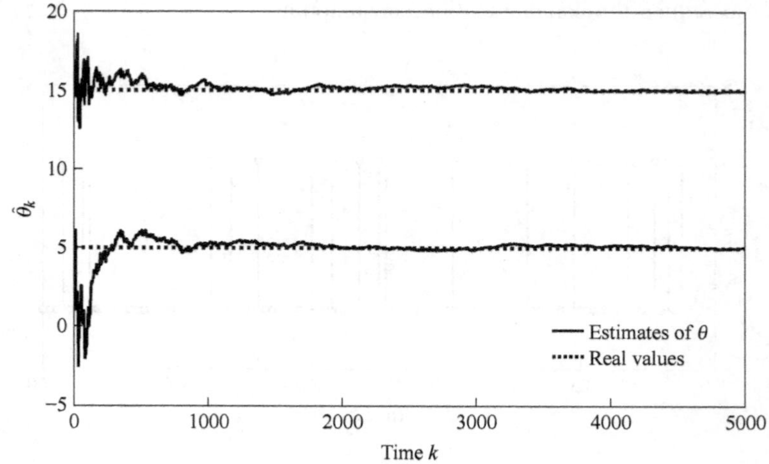

Fig. 4. 4 Convergence of $\hat{\theta}_k$

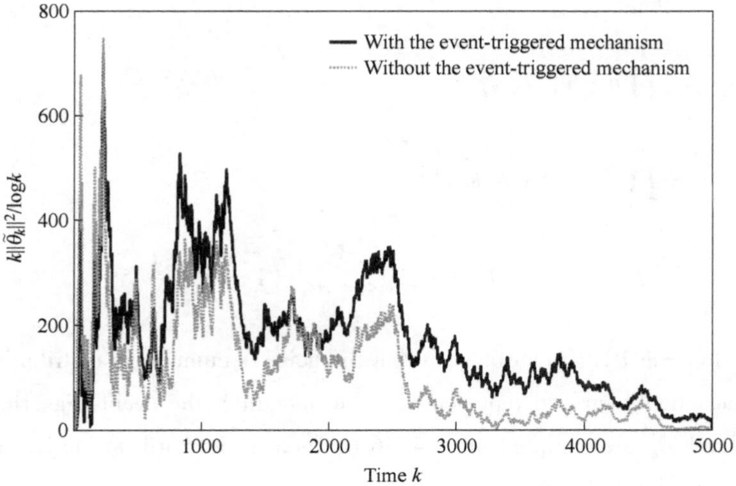

Fig. 4.5 Convergence rate shown by $k \| \tilde{\theta}_k \|^2 / \log k$

4.5 Notes

Saving the communication resource is a huge challenge in the current information era. The event-triggered scheme is a kind of frequently used techniques. This chapter studies it on the identification of FIR systems with binary-valued observations. A recursive algorithm is proposed to estimate the unknown parameter. Under some mild conditions, it is shown that the algorithm can strongly converge to the real value. The convergence rate of the estimation error is obtained. Also the communication rate is discussed.

Different from the scheduled measurement in Chapter 3 and the controlled communication, τ_k and δ_k in this chapter can be any $\mathcal{F}_{k-1}$-measurable functions. At time k, the information accessed by the estimator is $z_k = \{\gamma_k s_k, \gamma_k\}$. The joint distribution function of $z_1, \cdots, z_k$ is denoted by $p_\theta(z_1, \cdots, z_k)$. Since $z_1, \cdots, z_k$ are not independent, it is known that

$$p_\theta(z_1, \cdots, z_k) \neq \prod_{i=1}^{k} p_\theta(z_i)$$

By the conditional-probability-distribution formula, the likelihood function could be obtained

$$
\begin{aligned}
& p_\theta(z_1, \cdots, z_k) \\
& = \prod_{i=1}^{k} p_\theta(z_i \mid \mathcal{F}_{i-1}) \\
& = \prod_{i=1}^{k} \{ [\breve{F}_i^{s_i} (1 - \breve{F}_i)^{1-s_i}]^{\gamma_i} \times \\
& [I_{\{\tau_i - \delta_i < 1 < \tau_i + \delta_i\}} \breve{F}_i + I_{\{\tau_i - \delta_i < 0 < \tau_i + \delta_i\}} (1 - \breve{F}_i)]^{1-\gamma_i} \}
\end{aligned}
\tag{4.36}
$$

where $\breve{F}_i = F(C - \phi_i^{\mathrm{T}}\theta)$. It can be seen that the noise's cumulative distribution function $F(\cdot)$ (usually nonlinear and represented by an integral), the event-triggering indicator γ_k, and τ_k and δ_k are coupled in (4.36). Then it is hard to derive an explicit expression to maximize (4.36).

The Expectation Maximization Algorithm, Quasi-Newton method and others may provide effective ways to solve (4.36). Different from the Maximum Likelihood-based estimation, this chapter takes SA way to address the problem. As we know, the SA approach is to construct directly algorithms and then discuss the convergence properties without a cost function or an estimation criterion, and the SA-type algorithms often have simple and clear forms and are easier to implement and use for applications.

5 Prediction-Based Identification of Quantized-Input FIR Systems with Quantized Observations

Mathematically, for a given system input sequence, a parameter estimation algorithm is a mapping from the observation sequence $\{z_k : k \in \mathcal{K}\}$ to the set of possible values of the unknown parameter (denoted by $\mathcal{O}$), which can be described by

$$\mathcal{K} \times \mathcal{Z} \to \mathcal{O}$$

where $\mathcal{Z}$ represents the value domain of z_k. In practice, the complexities of $\mathcal{K}$ and $\mathcal{Z}$ usually determine the resource of storage space, transmission bandwidth and energy consumption and so on. Within the allowable range of estimation accuracy, it is very important and interesting to select smaller complexities of $\mathcal{K}$ and $\mathcal{Z}$, to reduce the burden of communication, the consumption of computation and the cost of memory.

The event-triggered communication scheme is an effective and frequently-used way to pick out the desired measured data, which is sampled only when a designed "event" occurs, and hence can simplify the measurement index set $\mathcal{K}$ while guaranteeing the system performance. The quantization is a process from a large set (such as the set of all real numbers) to a smaller set (often a finite set of discrete values), and hence can greatly simplify the observation value domain $\mathcal{Z}$. This can be directly grasped from the case of binary-valued quantized observation, where $\mathcal{Z} = \{0, 1\}$.

For simplifying both $\mathcal{K}$ and $\mathcal{Z}$, an intuitive way is the integrated use of the event-triggered scheme and the quantization. This chapter proposes a prediction-based event-triggered communication scheme to investigate the identification of FIR systems with quantized observations. For the binary-valued output quantization, an event-triggered communication scheme is introduced. Combining the empirical-measure-based identification technique and the weighted least-squares optimization, an identification algorithm is proposed to estimate the unknown parameter by making the most of the received data, the triggered indicator, the quantization threshold, and the statistical property of the system noise. Under quantized inputs, the algorithm is proved to be strongly convergent and its mean-square convergence rate and asymptotic efficiency is also established in terms of the CRLB. The limit of the average communication rate is derived and its some properties are given. The tradeoff between the communication cost and the estimation performance is formulated to a constrained minimization problem and discussed. The method and results are extended to the case of multi-threshold quantized observations.

The remaining parts of the chapter are arranged into the following sections. Section 5. 1 describes the prediction-based identification problem with binary-valued observations under quantized inputs. Section 5. 2 designs the parameter estimation algorithm and establishes the strong convergence and the asymptotic efficiency of the algorithm. Section 5. 3 discusses the tradeoff between the communication cost and the estimation performance. Section 5. 4 considers the case of multi-threshold quantized observations. Section 5. 5 simulates a numerical example to demonstrate the effectiveness of the algorithm and the main results obtained.

5. 1 Problem Formulation

Consider an SISO FIR system

$$y_k = a_1 u_k + \cdots + a_n u_{k-n+1} + d_k = \boldsymbol{\phi}_k^{\mathrm{T}} \boldsymbol{\theta} + d_k,\ k = 1,2,\cdots, \tag{5.1}$$

where $\boldsymbol{\phi}_k = [u_k, \cdots, u_{k-n+1}]^{\mathrm{T}}$, $\boldsymbol{\theta} = [a_1, \cdots, a_n]^{\mathrm{T}} \in \mathbf{R}^n$ is the unknown parameters, and d_k is the system noise.

As shown in Figure 5. 1, the input is finitely quantized with r possible values, that is, $u_k \in \mathcal{U} = \{\mu_1, \cdots, \mu_r\}$, which makes that the sequence $\{\phi_k^{\mathrm{T}}\}$ can only take values in $l = r^n$ possible (row vector) patterns denoted by

$$\mathcal{P} = \{\pi_1, \cdots, \pi_l\}$$

The output y_k is measured by a binary-valued sensor with threshold $C \in (-\infty, +\infty)$, which can be represented by

$$s_k = I_{\{y_k \leq C\}} \tag{5.2}$$

A prediction-based detector is employed to decide whether the binary-valued output observation s_k is transmitted to the EC, and it can also access $\hat{\theta}_{k-1}$ that is either broadcasted by the EC or computed by itself, where $\hat{\theta}_{k-1}$ denotes the estimate of θ at time $k-1$.

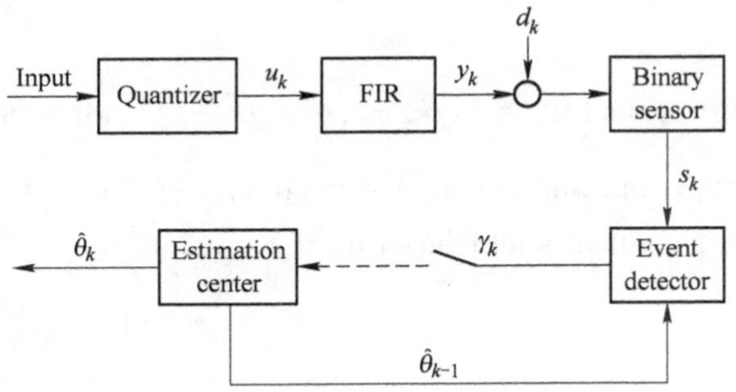

Fig. 5. 1 System set-up

γ_k is given by

$$\gamma_k = \begin{cases} 1, & s_k \neq \hat{s}_k \\ 0, & s_k = \hat{s}_k \end{cases} \tag{5.3}$$

where $\hat{s}_k = I_{\{\phi_k^{\mathrm{T}}\hat{\theta}_{k-1} \leqslant C\}}$ can be seen as the prediction of s_k based on the estimate $\hat{\theta}_{k-1}$.

We will first construct an estimation algorithm for θ and then establish the convergence properties. The limit of the average communication rate will be obtained, and the tradeoff between it and the estimation quantity will be discussed.

5. 2 Identification Algorithm and Convergence Performance

Let $\mathbb{K}_j = \{i: 1 \leqslant i \leqslant k, \phi_i^{\mathrm{T}} = \pi_j\}$ and k_j be the number of elements of $\mathbb{K}_j$, i. e. ,

$$k_j = \sum_{i=1}^{k} I_{\{\phi_i^{\mathrm{T}} = \pi_j\}}, \ j \in L \tag{5.4}$$

That is, $\{\phi_1^{\mathrm{T}}, \cdots, \phi_k^{\mathrm{T}}\}$ contains k_j copies of the pattern π_j.

Assumption 5. 1 The input sequence u is a deterministic signal and there exists $\beta_j \geqslant 0$ such that $\lim\limits_{k\to\infty} k_j/k = \beta_j$, $j \in L$. Without loss of generality, suppose that $\beta_j \neq 0$ for $j \in L_0$ and $\beta_j = 0$ for $j = L_0^-$.

Definition 5. 1 The pattern π_j is said to be persistent if $\beta_j > 0$. The input u is said to be *persistently exciting* if the matrix

$$\boldsymbol{\Psi} = \begin{bmatrix} \pi_1 \\ \vdots \\ \pi_{l_0} \end{bmatrix} \in \mathbf{R}^{l_0 \times n} \tag{5.5}$$

is full column rank.

Denote $j_k^0 = \sum_{j=1}^{l} j I_{\{\phi_k^T = \pi_j\}}$ and $\Psi_k = [\sqrt{k_1}\pi_1^T, \cdots, \sqrt{k_l}\pi_l^T]^T$, where Ψ is from (5.5). Supposed that $\Psi_k^T \Lambda \Psi_k$ is full rank, where $\Lambda = \mathrm{diag}(\lambda_1, \cdots, \lambda_l) > 0$ is a weighting matrix, an identification algorithm is introduced by

$$\eta_k = \gamma_k s_k + (1 - \gamma_k)\hat{s}_k \tag{5.6}$$

$$\xi_{k,j} = \begin{cases} \xi_{k-1,j}, & j \neq j_k^0 \\ \left(1 - \dfrac{1}{k_j}\right)\xi_{k-1,j} + \dfrac{1}{k_j}\eta_k, & j = j_k^0 \end{cases} \tag{5.7}$$

$$\hat{W}_k = [\sqrt{k_1}(C - F^{-1}(\xi_{k,1})), \cdots, \sqrt{k_l}(C - F^{-1}(\xi_{k,l}))]^T \tag{5.8}$$

$$\hat{\theta}_k = \left(\frac{1}{k}\Psi_k^T \Lambda \Psi_k\right)^{-1} \frac{1}{k}\Psi_k^T \Lambda \hat{W}_k \tag{5.9}$$

where the initial value $\xi_{0,j} \in (0,1)$ for $j \in L$, C is the threshold in (5.2) and $F(\cdot)$ is the distribution function of d_k.

Represent the covariance matrix of the estimation error by Σ_k, i. e.

$$\Sigma_k = E(\hat{\theta}_k - \theta)(\hat{\theta}_k - \theta)^T$$

and let $\Xi^* = \mathrm{diag}(\beta_1\rho_1^{-2}, \cdots, \beta_{l_0}\rho_{l_0}^{-2})$, where $\rho_j^2 = \varrho(C - \pi_j\theta)$ for $j \in L$, and $\varrho(\cdot)$ is from (2.2).

Theorem 5.1 Consider system (5.1) with binary-valued observations (5.2) and triggering mechanism (5.3) under Assumption 5.1. If the system noise $\{d_k\}$ follows Assumption 2.1 and the input u is persistently exciting, then

(1) $\hat{\theta}_k$ from the algorithm (5.6) ~ (5.9) converges strongly to the true value θ, i. e., $\hat{\theta}_k \to \theta$, w. p. 1, as $k \to \infty$.

(2) $k\Sigma_k \to (\Psi^T \Xi_1 \Psi)^{-1} \Psi^T \Xi_2 \Psi (\Psi^T \Xi_1 \Psi)^{-1}$ as $k \to \infty$, where

$$\Xi_1 = \mathrm{diag}(\lambda_1\beta_1, \cdots, \lambda_{l_0}\beta_{l_0}) \tag{5.10}$$

$$\Xi_2 = \mathrm{diag}(\beta_1\lambda_1^2\rho_1^2, \cdots, \beta_{l_0}\lambda_{l_0}^2\rho_{l_0}^2) \tag{5.11}$$

Proof: If $\gamma_k = 1$ then by (5.6) it follows that $\eta_k = \gamma_k s_k + (1-\gamma_k)\hat{s}_k = s_k$. If $\gamma_k = 0$, then from (5.3) it is known that $\hat{s}_k = s_k$, and then $\eta_k = \gamma_k s_k + (1-\gamma_k)\hat{s}_k = s_k$. In conclusion, we always have $\eta_k = s_k$, which together with (5.7) indicates that

$$\xi_{k,j} = \frac{1}{k_j}\sum_{i\in\mathbb{K}_j} s_i$$

This implies that (5.9) is just the algorithm (2.10) for the case $m=1$ (N is replaced by k), by Theorem 2.5 and Theorem 2.7 the proof follows. □

Lemma 5.1 The CRLB for estimating θ based on $\{s_1,\cdots,s_k\}$ is

$$\Sigma_k^{CR} = \left(\sum_{j=1}^{l} k_j \pi_j^{\mathrm{T}} \pi_j \rho_j^{-2}\right)^{-1}$$

where k_j is given by (5.4) for $j\in L$.

Proof: In Theorem 2.9, we let $m=1$ and the lemma can be given. □

Theorem 5.2 Under the condition of Theorem 5.1, if the weighting matrix Λ in (5.9) is selected as $\mathrm{diag}(\rho_1^{-2},\cdots,\rho_{l_0}^{-2},\lambda_{l_0+1},\cdots,\lambda_l)$, then the estimate $\hat{\theta}_k$ is asymptotically efficient in the sense that

$$k(\Sigma_k - \Sigma_k^{CR}) \to 0 \quad \text{as} \quad k\to\infty$$

Proof: Note that (5.10) and (5.11). Since $\lambda_j = \rho_j^{-2}$ for $j\in L_0$, it can be verified that $\Xi_1 = \Xi_2 = \Xi^*$ and $(\Psi^{\mathrm{T}}\Xi_1\Psi)^{-1}\Psi^{\mathrm{T}}\Xi_2\Psi(\Psi^{\mathrm{T}}\Xi_1\Psi)^{-1} = (\Psi^{\mathrm{T}}\Xi^*\Psi)^{-1}$. By virtue of Theorem 5.1, we have

$$k\Sigma_k \to (\Psi^{\mathrm{T}}\Xi^*\Psi)^{-1} \quad \text{as} \quad k\to\infty \tag{5.12}$$

By Assumption 5.1, it is known that $\beta_j = 0$ for $j = l_0+1,\cdots,l$. According to Lemma 5.1, it can be seen that

$$\begin{aligned} k\Sigma_k^{CR} &= \left(\sum_{j=1}^{l} \frac{k_j}{k}\pi_j^{\mathrm{T}}\pi_j\rho_j^{-2}\right)^{-1} \\ &\to \left(\sum_{j=1}^{l}\beta_j\pi_j^{\mathrm{T}}\pi_j\rho_j^{-2}\right)^{-1} = \left(\sum_{j=1}^{l_0}\beta_j\pi_j^{\mathrm{T}}\pi_j\rho_j^{-2}\right)^{-1} \\ &= (\Psi^{\mathrm{T}}\Xi^*\Psi)^{-1} \quad \text{as} \quad k\to\infty \end{aligned}$$

which together with (5.12) completes the proof. □

5.3 Tradeoff Between the Estimation Performance and the Communication Cost

Since $\lim_{k\to\infty} k\Sigma_k = \lim_{k\to\infty} k\Sigma_k^{CR} = (\Psi^T\Xi^*\Psi)^{-1}$ by Theorem 5.2, we can use $(\Psi^T\Xi^*\Psi)^{-1}$ as the estimation performance index. By the definition of Ξ^*, it can be seen that

$$\Xi^* = \mathrm{diag}(\beta_1\varrho^{-1}(C-\pi_1\theta),\cdots,\beta_{l_0}\varrho^{-1}(C-\pi_{l_0}\theta))$$

which indicates that Ξ^* is a function of C and one can denote $\Xi^* = \Xi^*(C)$.

Define the average communication rate of the event trigger (5.3) as

$$\bar{\gamma}_k = \frac{1}{k}\sum_{i=1}^{k}\gamma_i,\quad k=1,2,\cdots \tag{5.13}$$

To a certain extent, the limit of $\bar{\gamma}_k$ can reveal the ability of (5.3) in saving the communication resource, and be employed as the communication cost index. The coming will give it. For convenience, let

$$\tilde{F}(z) = I_{\{z<0\}}F(z) + I_{\{z\geqslant 0\}}(1-F(z)),\quad z\in\mathbf{R} \tag{5.14}$$

where $F(\cdot)$ is the distribution function of the system noise.

Theorem 5.3 Under the condition of Theorem 5.1, the average communication rate from the event trigger (5.3) is convergent, i.e.

$$\bar{\gamma}_k \to \bar{\gamma}(C) = \sum_{j=1}^{l_0}\beta_j\tilde{F}(C-\pi_j\theta),\text{w. p. 1.}\quad \text{as}\quad k\to\infty \tag{5.15}$$

Proof: Let $\mathcal{F}_{k-1}$ be the σ-algebra generated by $d_1,\cdots,d_{k-1}$, i.e. $\mathcal{F}_{k-1} = \sigma(d_i: 1\leqslant i\leqslant k-1)$. Owing to (5.3), it can be seen that

$$\begin{aligned}\hat{\gamma}_k &= \Pr(\gamma_k = 1 \mid \mathcal{F}_{k-1})\\ &= \Pr(s_k \neq \hat{s}_k \mid \mathcal{F}_{k-1})\\ &= \Pr(\hat{s}_k \neq 1 \mid \mathcal{F}_{k-1})\Pr(s_k = 1) + \Pr(\hat{s}_k \neq 0 \mid \mathcal{F}_{k-1})\Pr(s_k = 0)\end{aligned}$$

$$= I_{\{\phi_k^T \hat{\theta}_{k-1} > C\}} F(C - \phi_k^T \theta) + I_{\{\phi_k^T \hat{\theta}_{k-1} \leqslant C\}} (1 - F(C - \phi_k^T \theta))$$

This implies that $E\{\gamma_k - \hat{\gamma}_k \mid \mathcal{F}_{k-1}\} = 0$. Therefore, $\{\gamma_k - \hat{\gamma}_k, \mathcal{F}_k, k \geqslant 1\}$ is an MDS (martingale difference sequence, see Appendix A. 1. 2).

Note that $\gamma_k \leqslant 1$, $\hat{\gamma}_k \leqslant 1$ and then $\sum_{k=1}^{\infty} \frac{E(\gamma_k - \hat{\gamma}_k)^2}{k^2} \leqslant 4 \sum_{k=1}^{\infty} \frac{1}{k^2} < \infty$. By Lemma A. 5, we have

$$\frac{1}{k} \sum_{i=1}^{k} (\gamma_i - \hat{\gamma}_i) \to 0, \text{w. p. 1} \quad \text{as} \quad k \to \infty \tag{5.16}$$

By virtue of Theorem 5. 1 and with $\hat{\gamma}_k^0 = I_{\{\phi_k^T \theta > C\}} F(C - \phi_k^T \theta) + I_{\{\phi_k^T \theta \leqslant C\}} (1 - F(C - \phi_k^T \theta)) = \tilde{F}(C - \phi_k^T \theta)$ by (5. 14), one can have $\hat{\gamma}_k - \hat{\gamma}_k^0 \to 0$ as $k \to \infty$, which and (5. 16) give that

$$\frac{1}{k} \sum_{i=1}^{k} (\gamma_i - \hat{\gamma}_i^0) \to 0, \text{w. p. 1} \quad \text{as} \quad k \to \infty \tag{5.17}$$

In view of (5. 4), it can be seen that

$$\begin{aligned} \sum_{i=1}^{k} \hat{\gamma}_i^0 &= \sum_{i=1}^{k} \sum_{j=1}^{l} I_{\{\phi_i^T = \pi_j\}} \tilde{F}(C - \pi_j \theta) \\ &= \sum_{j=1}^{l} \Big(\sum_{i=1}^{k} I_{\{\phi_i^T = \pi_j\}} \Big) \tilde{F}(C - \pi_j \theta) \\ &= \sum_{j=1}^{l} k_j \tilde{F}(C - \pi_j \theta) \end{aligned}$$

By Assumption 5. 1, it follows that

$$\frac{1}{k} \sum_{i=1}^{k} \hat{\gamma}_i^0 = \sum_{j=1}^{l} \frac{k_j}{k} \tilde{F}(C - \pi_j \theta) \to \sum_{j=1}^{l} \beta_j \tilde{F}(C - \pi_j \theta) \quad \text{as} \quad k \to \infty$$

From (5. 13) and (5. 17), the proof can be obtained by the above. □

Proposition 5. 1 Under the condition of Theorem 5. 1, $\bar{\gamma}(C)$ is monotonically increasing on $(-\infty, \min_{1\leqslant j\leqslant l_0}\pi_j\theta]$ and monotonically decreasing on $[\max_{1\leqslant j\leqslant l_0}\pi_j\theta, \infty)$ with respect to C.

Proof: Since $F(z)$ is the accumulative distribution function of d_1, $F(z)$ is monotonically increasing on $(-\infty, \infty)$. In view of (5.14), it is known that

$$\tilde{F}(z) = \begin{cases} F(z), & z < 0 \\ 1 - F(z), & z \geqslant 0 \end{cases} \tag{5.18}$$

and hence from (5.15) one can have

$$\bar{\gamma}(C) = \begin{cases} \sum_{j=1}^{l_0} \beta_j F(C - \pi_j\theta), & C \leqslant \min_{1\leqslant j\leqslant l_0} \pi_j\theta \\ \sum_{j=1}^{l_0} \beta_j (1 - F(C - \pi_j\theta)), & C \geqslant \max_{1\leqslant j\leqslant l_0} \pi_j\theta \end{cases}$$

Due to $\beta_j > 0$ for $j \in L_0$, the proof follows. □

Example 1 Suppose $\theta = [-2, 5]^{\mathrm{T}}$. The noise is normally distribution with mean zero and standard deviation $\sigma = 5$. The quantized input makes that $\pi_1 = [-1, -1]$, $\pi_2 = [-1, 2]$, $\pi_3 = [2, -1]$, $\pi_4 = [2, 2]$ and $\beta_1 = 0.1$, $\beta_2 = 0.3$, $\beta_3 = 0.3$, $\beta_4 = 0.3$ with $l_0 = 4$, which implies that $\max_{1\leqslant j\leqslant l_0} \pi_j\theta = 12$, $\min_{1\leqslant j\leqslant l_0}\pi_j\theta = -9$. By (5.15), we have

$$\bar{\gamma}(C) = 0.1\tilde{F}(C + 3) + 0.3\tilde{F}(C - 12) + 0.3\tilde{F}(C + 9) + 0.3\tilde{F}(C - 6)$$

whose graph is given by Figure 5.2. One can see that $\bar{\gamma}(C)$ is indeed monotonically decreasing on $[12, \infty)$ and monotonically increasing on $(-\infty, -9]$.

In Example 1, we have seen that $\bar{\gamma}(C) \to 0$ as $C \to \infty$ or $C \to -\infty$. With

$$\boldsymbol{\Psi} = \begin{bmatrix} -1 & -1 \\ -1 & 2 \\ 2 & -1 \\ 2 & 2 \end{bmatrix}$$

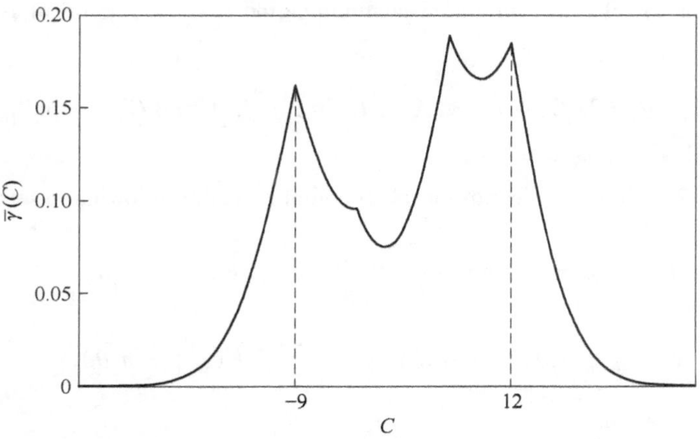

Fig. 5.2 The graph of $\overline{\gamma}(C)$ with respect to C

Figure 5.3 shows the graph of $\|(\Psi^{\mathrm{T}}\Xi^{*}(C)\Psi)^{-1}\|$, where $\|\cdot\|$ denotes the Frobenius Norm and one can see that $\|(\Psi^{\mathrm{T}}\Xi^{*}(C)\Psi)^{-1}\| \to \infty$ as $C \to \infty$ or $C \to -\infty$.

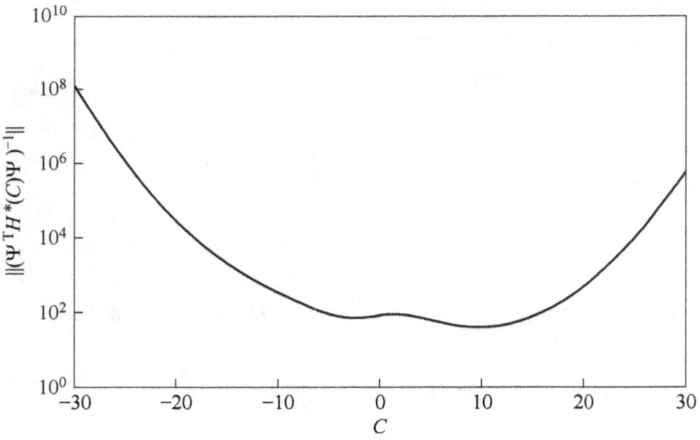

Fig. 5.3 The graph of $\|(\Psi^{\mathrm{T}}\Xi^{*}(C)\Psi)^{-1}\|$ with respect to C

Then, an interesting issue is how to balance the estimation performance and the communication cost. This may be stated as a constrained minimization problem

$$\min_{C}\overline{\gamma}(C)$$

$$\text{s.t. } (\Psi^{\mathrm{T}}\Xi^{*}(C)\Psi)^{-1} \leqslant \Delta \tag{5.19}$$

where $\Delta > 0$ is a given positive definite matrix. That is, for a given estimation accuracy

Δ ,(5. 19) aims to solve the minimum communication cost. In general, it is hard to obtain its explicit solution.

Let $\mathcal{D}_\Delta = \{C: (\Psi^T \Xi^*(C)\Psi)^{-1} \leqslant \Delta\}$, $\underline{C}_\Delta = \inf_{C\in\mathcal{D}_\Delta} C$ and $\overline{C}_\Delta = \sup_{C\in\mathcal{D}_\Delta} C$. Then we have the following proposition.

Proposition 5. 2 Under the condition of Theorem 5. 1, the following assertions hold.

(1) If $\underline{C}_\Delta \geqslant \max_{1\leqslant j\leqslant l_0} \pi_j\theta$, then $\inf_{C\in\mathcal{D}_\Delta} \overline{\gamma}(C) = \sum_{j=1}^{l_0} \beta_j \tilde{F}(\overline{C}_\Delta - \pi_j\theta)$.

(2) If $\overline{C}_\Delta \leqslant \min_{1\leqslant j\leqslant l_0} \pi_j\theta$, then $\inf_{C\in\mathcal{D}_\Delta} \overline{\gamma}(C) = \sum_{j=1}^{l_0} \beta_j \tilde{F}(\underline{C}_\Delta - \pi_j\theta)$.

(3) In general, we have

$$\inf_{C\in\mathcal{D}_\Delta} \overline{\gamma}(C) \geqslant \sum_{j=1}^{l_0} \beta_j \min\{\tilde{F}(\underline{C}_\Delta - \pi_j\theta), \tilde{F}(\overline{C}_\Delta - \pi_j\theta)\}$$

Proof: (1) If $\underline{C}_\Delta \geqslant \max_{1\leqslant j\leqslant l_0} \pi_j\theta$, then, according to Proposition 5. 1, it is known that $\overline{\gamma}(C)$ is monotonically decreasing on $[\underline{C}_\Delta, \overline{C}_\Delta]$, which gives the desired result.

(2) The proof is similar to the one of (1).

(3) By (5. 15), it is known that

$$\inf_{C\in\mathcal{D}_\Delta} \overline{\gamma}(C) \geqslant \sum_{j=1}^{l_0} \beta_j \inf_{C\in D_\Delta} \tilde{F}(C - \pi_j\theta) \tag{5.20}$$

On account of (5. 18), we have

$$\inf_{C\in\mathcal{D}_\Delta} \tilde{F}(C - \pi_j\theta) = \min\{\tilde{F}(\underline{C}_\Delta - \pi_j\theta), \tilde{F}(\overline{C}_\Delta - \pi_j\theta)\}$$

which and (5. 20) complete the proof. □

5. 4 Multi-Threshold Quantized Observations

This section considers the case of multi-threshold quantized output observations, where y_k is measured by a sensor of m thresholds $-\infty < C_1 < \cdots < C_m < \infty$. With $C_0 = -\infty$ and $C_{m+1} = \infty$, the observation sensor can be represented by

$$s_k = \sum_{i=1}^{m+1} i I_{\{y_k \in (C_{i-1}, C_i]\}} \tag{5.21}$$

Hence, $s_k = i$ implies that $y_k \in (C_{i-1}, C_i]$ for $i = 1, \cdots, m+1$. An alternative representation of (5.21) is to view the multi-threshold quantized observation as a vector-valued binary observation in which each vector component represents the output of one threshold, by defining

$$\tilde{s}_k = [s_k^1, \cdots, s_k^m]^T$$

where $$s_k^i = I_{\{-\infty < y_k \leqslant C_i\}}, \quad i = 1, \cdots, m \tag{5.22}$$

The triggering condition γ_k is given by

$$\gamma_k = \begin{cases} 1, & s_k \neq \hat{s}_k \\ 0, & s_k = \hat{s}_k \end{cases} \tag{5.23}$$

$$\hat{s}_k = \sum_{i=1}^{m+1} i I_{\{\phi_k^T \hat{\theta}_{k-1} \in (C_{i-1}, C_i]\}} \tag{5.24}$$

Let $\gamma_k s_k + (1 - \gamma_k)\hat{s}_k = \eta_k$ that is available for algorithm design, and

$$\eta_k^i = \begin{cases} 1, & \eta_k \leqslant i \\ 0, & \text{otherwise} \end{cases} \quad i = 1, \cdots, m$$

Then it can be seen that $[\eta_k^1, \cdots, \eta_k^m]^T = \tilde{s}_k$ by (5.22) ~ (5.24). As a result, we can use $\eta_k^1, \cdots, \eta_k^m$ to construct m algorithms as (5.6) ~ (5.9), and obtain m estimators of θ, denoted by $\hat{\theta}_k, \cdots, \hat{\theta}_k^m$. Define $\nu = [\nu_1, \cdots, \nu_m]^T$ such that $\nu_1 + \cdots + \nu_m = 1$. One can construct an estimator of θ by

$$\hat{\theta}_k = \sum_{i=1}^{m} \nu_i \hat{\theta}_k^i$$

which is the QCCE proposed by [39]. Under quantized inputs, the corresponding convergence performance and the asymptotic efficiency can be given as in [41]. In order to avoid unnecessary duplication, we omit them.

The limit of the average communication rate from (5.23) is established by the following theorem.

Theorem 5.4 Consider system (5.1) with quantized observations (5.21) and triggering mechanism (5.23) under Assumptions 2.1 and 5.1. If the input u is persistently exciting, then $\overline{\gamma}_k$ from (5.23) can converge to

$$\overline{\gamma} = \sum_{j=1}^{l_0} \beta_j \sum_{i=1}^{m+1} \left(I_{\{\pi_j \theta \notin (C_{i-1}, C_i]\}} \times (F(C_i - \pi_j \theta) - F(C_{i-1} - \pi_j \theta)) \right), \text{w. p. } 1$$

Proof: Define $\hat{\gamma}_k^0 = \sum_{i=1}^{m+1} I_{\{\phi_k^{\mathrm{T}} \theta \notin (C_{i-1}, C_i]\}} (F(C_i - \phi_k^{\mathrm{T}} \theta) - F(C_{i-1} - \phi_k^{\mathrm{T}} \theta))$. By (5.4), one can have

$$\sum_{i=1}^{k} \hat{\gamma}_i^0 = \sum_{i=1}^{k} \sum_{j=1}^{l} I_{\{\phi_k^{\mathrm{T}} = \pi_j\}} \sum_{i=1}^{m+1} \left(I_{\{\pi_j \theta \notin (C_{i-1}, C_i]\}} \times (F(C_i - \pi_j \theta) - F(C_{i-1} - \pi_j \theta)) \right)$$

$$= \sum_{j=1}^{l} k_j \sum_{i=1}^{m+1} \left(I_{\{\pi_j \theta \notin (C_{i-1}, C_i]\}} \times (F(C_i - \pi_j \theta) - F(C_{i-1} - \pi_j \theta)) \right)$$

which together with Assumption 5.1 yields that

$$\frac{1}{k} \sum_{i=1}^{k} \hat{\gamma}_i^0 = \sum_{j=1}^{l} \frac{k_j}{k} \sum_{i=1}^{m+1} \left(I_{\{\pi_j \theta \notin (C_{i-1}, C_i]\}} \times (F(C_i - \pi_j \theta) - F(C_{i-1} - \pi_j \theta)) \right)$$

$$\to \sum_{j=1}^{l_0} \beta_j \sum_{i=1}^{m+1} \left(I_{\{\pi_j \theta \notin (C_{i-1}, C_i]\}} \times (F(C_i - \pi_j \theta) - F(C_{i-1} - \pi_j \theta)) \right)$$

$$\text{w. p. } 1 \quad \text{as} \quad k \to \infty \tag{5.25}$$

By (5.1) and Assumption 2.1, we have

$$\Pr(s_k = i) = \Pr(C_{i-1} < y_k \leqslant C_i)$$

$$= \Pr(C_{i-1} < \phi_k^{\mathrm{T}} \theta + d_k \leqslant C_i)$$

$$= F(C_i - \phi_k^{\mathrm{T}} \theta) - F(C_{i-1} - \phi_k^{\mathrm{T}} \theta)$$

Noticing that $\hat{s}_k \neq i$ if and only if $\phi_k^{\mathrm{T}} \hat{\theta}_{k-1} \notin (C_{i-1}, C_i]$ by (5.24), it is known that

$$\hat{\gamma}_k = E[\gamma_k \mid \mathcal{F}_{k-1}] = \Pr(s_k \neq \hat{s}_k \mid \mathcal{F}_{k-1})$$

$$= \sum_{i=1}^{m+1} \Pr(\hat{s}_k \neq i \mid \mathcal{F}_{k-1}) \Pr(s_k = i)$$

$$= \sum_{i=1}^{m+1} I_{\{\hat{s}_k \neq i\}} (F(C_i - \phi_k^{\mathrm{T}}\theta) - F(C_{i-1} - \phi_k^{\mathrm{T}}\theta))$$

$$= \sum_{i=1}^{m+1} (I_{\{\phi_k^{\mathrm{T}}\hat{\theta}_{k-1} \notin (C_{i-1}, C_i]\}} \times (F(C_i - \phi_k^{\mathrm{T}}\theta) - F(C_{i-1} - \phi_k^{\mathrm{T}}\theta)))$$

and $E\{\gamma_k - \hat{\gamma}_k \mid \mathcal{F}_{k-1}\} = 0$. Therefore, $\{\gamma_k - \hat{\gamma}_k, \mathcal{F}_k, k \geq 1\}$ is an MDS. Recalling Lemma A. 5, we have

$$\frac{1}{k}\sum_{i=1}^{k} (\gamma_i - \hat{\gamma}_i) \to 0, \text{w. p. } 1 \quad \text{as} \quad k \to \infty \tag{5.26}$$

By the convergence of $\hat{\theta}_k$, it can be obtained that $\hat{\gamma}_k - \hat{\gamma}_k^0 \to 0$, w. p. 1 as $k \to \infty$, which together with (5. 25) and (5. 26) can give the proof. □

5.5 Numerical Simulation

Consider a gain system $y_k = u_k\theta + d_k$, where the true value $\theta = 3$ and $\{d_k\}$ is a sequence of i. i. d. normal random variables with zero mean and standard deviation $\sigma = 6$. The output is measured by a binary-valued sensor with the threshold $C = 2$. The input is quantized and takes value from $\mathcal{U} = \{\pi_1, \pi_2, \pi_3\} = \{-1.5, 2, 1\}$. Since $\theta \in \mathbf{R}$, we have $\mathcal{P} = \mathcal{U}$. At k, assume that

$$k_1 = k - k_2 - k_3, k_2 = 0.6(k - k_3), k_3 = \min\{110, |\log k|\}$$

Thus, it is known that $\beta_1 = 0.4, \beta_2 = 0.6$, $\pi_1 = -1.5$ and $\pi_2 = 2$ are persistent, and then $\Psi = [-1.5, 2]^{\mathrm{T}}$.

The event-triggered communication scheme (5. 3) and the algorithm (5. 6) ~ (5. 9) is used to estimate θ with $\Lambda = \mathrm{diag}(0.0114, 0.015, 0.001)$ and $\xi_{0,j} = 1/2$ for $j \in L$. The convergence is shown by Figure 5. 4. The average of 20 trajectories of $k(\hat{\theta}_k - \theta)(\hat{\theta}_k - \theta)^{\mathrm{T}}$ is employed to approximate $k\Sigma_k$, and the difference between it and $k\Sigma_k^{CR}$ is gradually getting smaller in Figure 5. 5, which illustrates the asymptotic efficiency of the algorithm.

The transmission on the time interval [200,250] is displayed in Figure 5.6, where only 12 measurements are send to EC. It can be calculated that

$$\bar{\gamma} = \sum_{j=1}^{l_0} \beta_j \tilde{F}(C - \pi_j \theta) = 0.2072$$

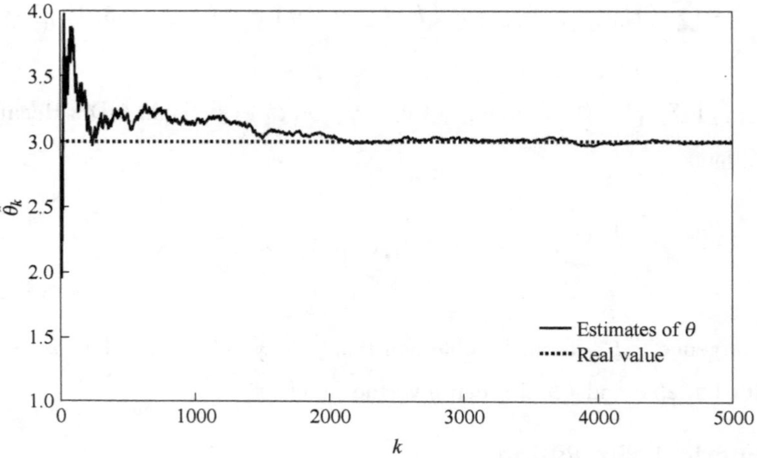

Fig. 5.4 Convergence of $\hat{\theta}_k$

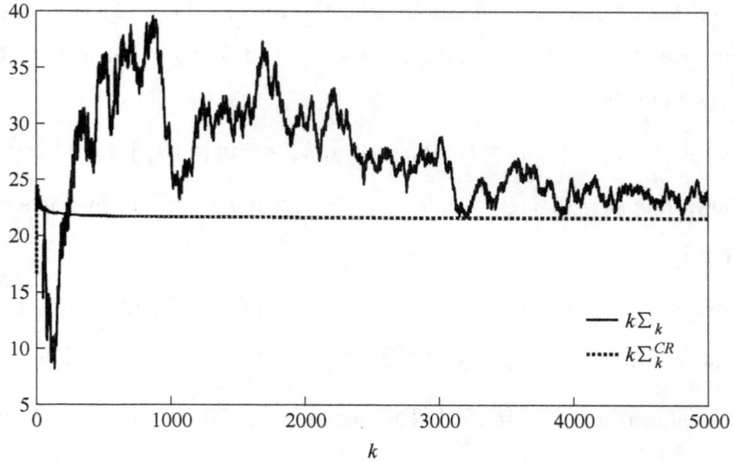

Fig. 5.5 Asymptotic efficiency of $\hat{\theta}_k$

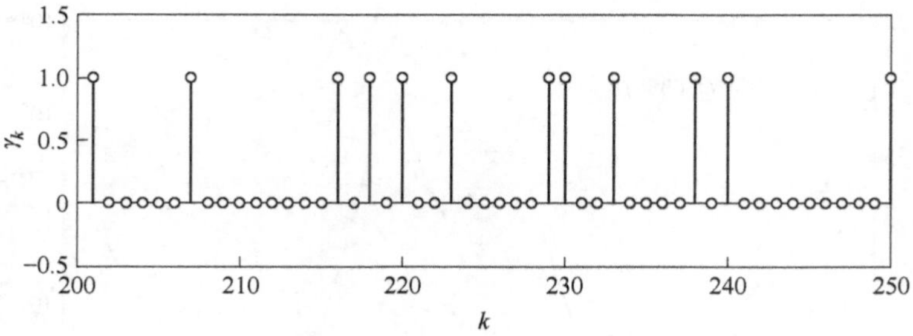

Fig. 5.6 Transmission on the time interval [200,250]

In Figure 5.7, we can see that $\overline{\gamma}_k \to 0.2072$ as $k \to \infty$, which is in accord with Theorem 5.3 by (5.13).

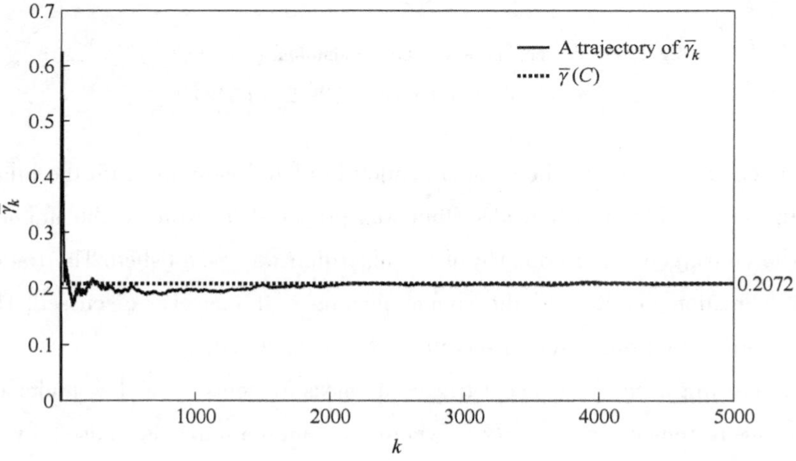

Fig. 5.7 Convergence of the average communication rate

Figure 5.8 shows the graphs of $\overline{\gamma}(C)$ and $(\Psi^{\mathrm{T}}\Xi^{*}(C)\Psi)^{-1}$ with respect to C, where the left vertical axis is for $\overline{\gamma}(C)$ and the right one is for $(\Psi^{\mathrm{T}}\Xi^{*}(C)\Psi)^{-1}$. Let the estimation accuracy requirement $\Delta = 30$. Then one can see that $\mathcal{D}_{\Delta} = [-3.1, 11.1]$, and the minimum value of $\overline{\gamma}(C)$ on $\mathcal{D}_{\Delta}$ is 0.1205.

5.6 Notes

It is interesting to employ the smallest complexity of the measurements to get desired system performance. For the identification of FIR systems with quantized inputs and quantized output observations, this chapter introduced a prediction-based event-triggered com-

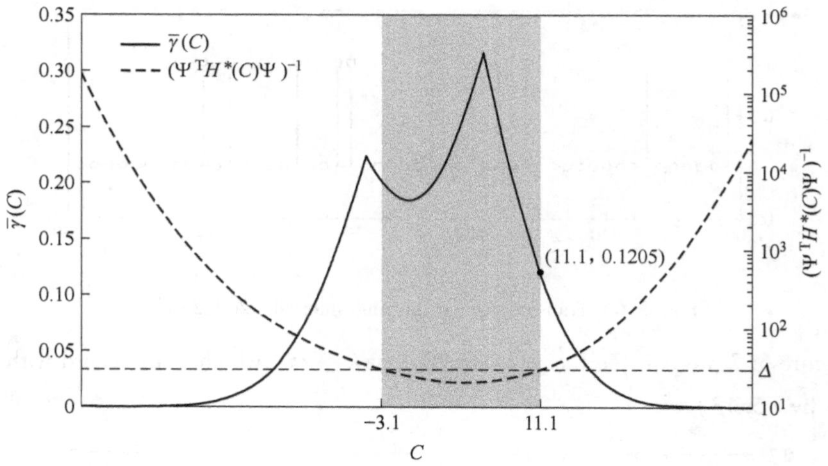

Fig. 5. 8 Tradeoff between the communication cost $\overline{\gamma}(C)$ and the estimation performance $(\Psi^{T}\Xi^{*}(C)\Psi)^{-1}$

munication schemes to reduce the communication burden. Beginning with the binary-valued quantization, an identification algorithm was proposed to estimate the unknown parameter. The convergence performance of the algorithm was established. The tradeoff between the estimation quality and the communication cost was also discussed. Then the case of multi-threshold quantized observations was considered.

It is interesting that the event-triggered measurements can be understood as intermittent measurements as in [74], where the intermittence is caused by random packet dropouts. It should be pointed out that the packet-dropout process and the system noise process are mutually independent, but the event-triggered measurements $\{\gamma_k s_k\}$ are correlated to the system noise process $\{d_k\}$. This is the essential difference between them.

In this chapter, the cumulative distribution function of the system noise is assumed to be known and employed to design identification algorithms. For the case that $F(\cdot)$ is unknown, one can refer to the methods in [17] and [41]

6 FIR System Identification under Either-or Communication with Quantized Inputs and Quantized Observations

Considering that the binary-valued quantization can only give two values, that just one of them is transmitted can still provide all the observation information since the other value is achieved at the not-transmitted time. Based on this idea and for a smaller communication rate, a so-called either-or communication scheme for the binary-valued quantization will be introduced in this chapter. For the multi-threshold quantization, it can be seen as a vector-valued observation in which each component represents the output of one threshold, that is, each component is a binary-valued observation, and then the either-or communication scheme can be applied to each component of the multi-threshold quantization.

Under the either-or communication scheme, we investigates the identification of FIR systems with quantized inputs and quantized output observations. By making the most of the received data, the triggered indicator and the not-triggered condition, an auxiliary sequence is constructed to restore the quantized observation. Together with the quantization threshold and the statistical property of the system noise, an identification algorithm is proposed to estimate the unknown parameter based on the empirical measure and weighted least squares. The convergence performance is established. By use of the strong law of large numbers of martingale difference sequences, the communication rate is derived and its some properties are given. Moreover, the case of multi-threshold quantized observations is considered.

The layout of the chapter is as follows. Section 6. 1 presents the quantized-input FIR system identification problem with both quantized and event-triggered output observations. Section 6. 2 introduces the either-or communication scheme and the parameter estimation algorithm. Section 6. 3 gives the convergence performance of the algorithm. Section 6. 4 deals with case of multi-threshold quantized observations. Section 6. 5 provides a numerical example to demonstrate the main results obtained.

6.1 Problem Formulation

Consider an SISO FIR system

$$y_k = a_1 u_k + \cdots + a_n u_{k-n+1} + d_k = \boldsymbol{\phi}_k^{\mathrm{T}} \boldsymbol{\theta} + d_k$$

$$k = 1, 2, \cdots \tag{6.1}$$

where $\boldsymbol{\phi}_k = [u_k, \cdots, u_{k-n+1}]^{\mathrm{T}}$, $\boldsymbol{\theta} = [a_1, \cdots, a_n]^{\mathrm{T}} \in \mathbf{R}^n$ is the unknown parameters, and d_k is the system noise.

The system set-up is shown by Figure 6.1, where the input is finitely quantized with r possible values. The output y_k is measured by a binary-valued sensor with the threshold $C \in (-\infty, +\infty)$, which can be represented by an indicator function

$$s_k = I_{\{y_k \leqslant C\}} \tag{6.2}$$

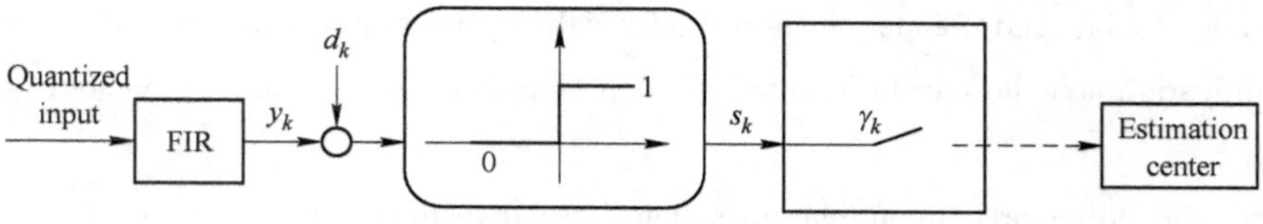

Fig. 6.1 System configuration

The event detector can access $\hat{\theta}_{k-1}$ that is either broadcasted by the EC or computed by itself, where $\hat{\theta}_{k-1}$ is the estimate of θ at time $k-1$. In general, the triggering condition may depend on ϕ_k, s_k, $\hat{\theta}_{k-1}$, and C. That is, γ_k is a function of ϕ_k, s_k, $\hat{\theta}_{k-1}$, and C.

This chapter will introduce a so-called either-or communication scheme, and construct an estimation algorithm for θ, and then establish the corresponding convergence properties. Moreover, the communication rate

$$\bar{\gamma} = \lim_{k \to \infty} \frac{1}{k} \sum_{i=1}^{k} \gamma_i$$

will be obtained, and the case of multi-threshold quantized observations will be discussed.

For convenience, define

$$\overline{F}(z) = \min\{F(z), 1 - F(z)\}, \quad z \in \mathbf{R} \tag{6.3}$$

Denote $\mathcal{X}_0 = F^{-1}\left(\frac{1}{2}\right)$. If $z \leqslant \mathcal{X}_0$, then it can be seen that $F(z) \leqslant \frac{1}{2}$, and then $1 - F(z) \geqslant \frac{1}{2}$, which implies that $1 - F(z) \geqslant F(z)$ and $\overline{F}(z) = F(z)$. If $z > \mathcal{X}_0$, then one can know that $F(z) \geqslant \frac{1}{2} \geqslant 1 - F(z)$, and hence $\overline{F}(z) = 1 - F(z)$. As a result, we have

$$\overline{F}(z) = \begin{cases} F(z), & z \leqslant \mathcal{X}_0 \\ 1 - F(z), & z > \mathcal{X}_0 \end{cases} \tag{6.4}$$

6.2 Either-or Communication Scheme and Identification Algorithm

Let $\mathbb{K}_j = \{i: 1 \leqslant i \leqslant k, \phi_i^{\mathrm{T}} = \pi_j\}$ and k_j be the number of elements of $\mathbb{K}_j$, i. e.

$$k_j = \sum_{i=1}^{k} I_{\{\phi_i^{\mathrm{T}} = \pi_j\}}, \quad j \in L \tag{6.5}$$

Assumption 6.1 There exists $\beta_j \geqslant 0$ such that $\lim\limits_{k\to\infty} k_j/k = \beta_j, j \in L$. The pattern π_j is said to be *persistent* if $\beta_j > 0$. Without loss of generality, suppose that $\beta_j \neq 0$ for $j \in L_0$ and $\beta_j = 0$ for $j = L_0^-$. The input u is persistently exciting, i. e. , the matrix

$$\boldsymbol{\Psi} = \begin{bmatrix} \pi_1 \\ \vdots \\ \pi_{l_0} \end{bmatrix} \in \mathbf{R}^{l_0 \times n}$$

is full column rank.

Since s_k can only be 1 or 0, we can still know the value of s_k at the not-triggered time if just one of them can be permitted to be sent to EC. For example, if only 1 is permitted to be sent (i. e. , $\gamma_k = 1$ if and only if $s_k = 1$), then $\gamma_k = 0$ means $s_k = 0$. Then, a key question is: which one is permitted to be sent can achieve a smaller communication rate?

If only 1 is permitted to be sent, that is, $\gamma_k = I_{\{s_k = 1\}}$, then it is known that

$$E\gamma_k = EI_{\{y_k \leqslant C\}} = F(C - \pi_j\theta) \quad \text{for} \quad \boldsymbol{\phi}_k^{\mathrm{T}} = \pi_j, \quad j \in L$$

and from Assumption 6. 1

$$\bar{\gamma} = \lim_{k\to\infty} \frac{1}{k} \sum_{i=1}^{k} \gamma_k = \lim_{k\to\infty} \frac{1}{k} \sum_{j=1}^{l} \sum_{i\in\mathbb{K}_j} \gamma_i$$

$$= \lim_{k\to\infty} \sum_{j=1}^{l} \frac{k_j}{k} \frac{1}{k_j} \sum_{i\in\mathbb{K}_j} \gamma_i = \sum_{j=1}^{l_0} \beta_j F(C - \pi_j\theta)$$

If 0 is done, i. e. , $\gamma_k = I_{\{s_k=0\}}$, then we have $E\gamma_k = 1 - F(C - \pi_j\theta)$ for $\phi_k = \pi_j, j \in L$, and

$$\bar{\gamma} = \sum_{j=1}^{l_0} \beta_j (1 - F(C - \pi_j\theta))$$

Notice (6. 4) and

$$\bar{F}(C - \pi_j\theta) = \begin{cases} F(C - \pi_j\theta), & C \leqslant \pi_j\theta + \chi_0 \\ 1 - F(C - \pi_j\theta), & C > \pi_j\theta + \chi_0 \end{cases}$$

For getting a smaller communication rate, γ_k should be $I_{\{C\leqslant\pi_j\theta+\chi_0, s_k=1\}} + I_{\{C>\pi_j\theta+\chi_0, s_k=0\}}$ for $\phi_k = \pi_j, j \in L$. This indicates

$$\gamma_k = \sum_{j=1}^{l} I_{\{\phi_k^{\mathrm{T}}=\pi_j\}} (I_{\{C\leqslant\pi_j\theta+\chi_0, s_k=1\}} + I_{\{C>\pi_j\theta+\chi_0, s_k=0\}}) \tag{6. 6}$$

which renders

$$\bar{\gamma} = \sum_{j=1}^{l_0} \beta_j \bar{F}(C - \pi_j\theta) \tag{6. 7}$$

However, θ is unknown and then (6. 6) is not implementable. We have to use an estimate of θ to substitute for its real value to yield an implementable event trigger as follows

$$\gamma_k = \sum_{j=1}^{l} I_{\{\phi_k^{\mathrm{T}}=\pi_j\}} (I_{\{C\leqslant\pi_j\hat{\theta}_{k-1}+\chi_0, s_k=1\}} + I_{\{C>\pi_j\hat{\theta}_{k-1}+\chi_0, s_k=0\}}) \tag{6. 8}$$

which is called either-or communication scheme. The next will construct an algorithm to generate $\hat{\theta}_k$.

Let $j_k^0 = \sum_{j=1}^{l} j I_{\{\phi_k^{\mathrm{T}}=\pi_j\}}$, $\Psi_k = [\sqrt{k_1}\pi_1^{\mathrm{T}}, \cdots, \sqrt{k_l}\pi_l^{\mathrm{T}}]^{\mathrm{T}}$, and suppose that $\Psi_k^{\mathrm{T}}\Lambda\Psi_k$ is full rank, where $\Lambda = \text{diag}(\lambda_1, \cdots, \lambda_l) > 0$ is a weighting matrix. An identification algorithm is introduced by

$$\eta_k = \sum_{j=1}^{l} I_{\{\phi_k^{\mathrm{T}} = \pi_j\}} I_{\{C > \pi_j \hat{\theta}_{k-1} + x_0\}} (1 - \gamma_k) \tag{6.9}$$

$$\xi_{k,j} = \begin{cases} \xi_{k-1,j}, & j \neq j_k^0 \\ \left(1 - \dfrac{1}{k_j}\right) \xi_{k-1,j} + \dfrac{1}{k_j}(\gamma_k s_k + \eta_k), & j = j_k^0 \end{cases} \tag{6.10}$$

$$\hat{W}_k = [\sqrt{k_1}(C - F^{-1}(\xi_{k,1})), \cdots, \sqrt{k_l}(C - F^{-1}(\xi_{k,l}))]^{\mathrm{T}} \tag{6.11}$$

$$\hat{\theta}_k = \left(\frac{1}{k}\Psi_k^{\mathrm{T}} \Lambda \Psi_k\right)^{-1} \frac{1}{k}\Psi_k^{\mathrm{T}} \Lambda \hat{W}_k \tag{6.12}$$

where the initial value $\xi_{0,j} \in (0,1)$ for $j \in L$.

6.3 Convergence Performance

This section gives the convergence performance of the identification algorithm, together with the communication rate and its some properties.

Theorem 6.1 Consider system (6.1) with binary-valued observations (6.2) and triggering mechanism (6.8) under Assumption 6.1. If the system noise follows Assumption 2.1, then

(1) $\hat{\theta}_k$ from the algorithm (6.9) ~ (6.12) converges strongly to the true value θ, i.e., $\hat{\theta}_k \to \theta$, $w.p.$ 1 as $k \to \infty$.

(2) $k\Sigma_k = kE(\hat{\theta}_k - \theta)(\hat{\theta}_k - \theta)^{\mathrm{T}} \to (\Psi^{\mathrm{T}}\Xi_1\Psi)^{-1}\Psi^{\mathrm{T}}\Xi_2\Psi(\Psi^{\mathrm{T}}\Xi_1\Psi)^{-1}$ as $k \to \infty$, where Ξ_1 and Ξ_2 are given by (5.10) and (5.11).

(3) if $\Lambda = \text{diag}(\rho_1^{-2}, \cdots, \rho_{l_0}^{-2}, \lambda_{l_0+1}, \cdots, \lambda_l)$, then $\hat{\theta}_k$ is asymptotically efficient in the sense that

$$k(\Sigma_k - \Sigma_k^{CR}) \to 0 \quad \text{as} \quad k \to \infty$$

where $\Sigma_k^{CR} = (\sum_{j=1}^{l} k_j \pi_j^{\mathrm{T}} \pi_j \rho_j^{-2})^{-1}$ is the CRLB for estimating θ based on $\{s_1, \cdots, s_k\}$ and

k_j is given by (6.5) for $j \in L$.

Proof: For any $k \geqslant 1$, there exists $j_k^* \in L$ such that $\phi_k^T = \pi_{j_k^*}$, and by (6.8) and (6.9) it is known that

$$\gamma_k = I_{\{C \leqslant \pi_{j_k^*} \hat{\theta}_{k-1} + \chi_0, s_k = 1\}} + I_{\{C > \pi_{j_k^*} \hat{\theta}_{k-1} + \chi_0, s_k = 0\}}$$

$$\eta_k = I_{\{C > \pi_{j_k^*} \hat{\theta}_{k-1} + \chi_0\}} (1 - \gamma_k)$$

According to the above, if $\gamma_k = 0$ and $s_k = 1$, then we have $C > \pi_{j_k^*} \hat{\theta}_{k-1} + \chi_0$ and $\eta_k = 1 - \gamma_k = s_k$. If $\gamma_k = 0$ and $s_k = 0$, then $C \leqslant \pi_{j_k^*} \hat{\theta}_{k-1} + \chi_0$ and $\eta_k = 0 = s_k$. For the case that $\gamma_k = 1$, it can be seen that $\eta_k = 0$ and $\gamma_k s_k + \eta_k = s_k$. In conclusion, we always have $\gamma_k s_k + \eta_k = s_k$, which together with (6.10) indicates that

$$\xi_{k,j} = \frac{1}{k_j} \sum_{i \in \mathbb{K}_j} s_i$$

This implies that (6.12) is just the identification algorithm (2.10) for the case of single threshold (N is replaced by k), and by Theorem 2.5, Theorem 2.7, Theorem 2.9, and Theorem 2.10 the proof can be obtained. □

Theorem 6.2 Under the condition of Theorem 6.1, the communication rate from the event trigger (6.8) is given by

$$\bar{\gamma} = \sum_{j=1}^{l_0} \beta_j \bar{F}(C - \pi_j \theta), \text{w. p. } 1 \tag{6.13}$$

Proof: Owing to (6.4) and (6.8), it can be seen that

$$\begin{aligned} \hat{\gamma}_k &= \Pr(\gamma_k = 1 \mid \mathcal{F}_{k-1}) \\ &= \sum_{j=1}^{l} I_{\{\phi_k^T = \pi_j\}} (I_{\{C > \pi_j \hat{\theta}_{k-1} + \chi_0\}} (1 - F(C - \pi_j \theta)) + I_{\{C \leqslant \pi_j \hat{\theta}_{k-1} + \chi_0\}} F(C - \pi_j \theta)) \end{aligned}$$

where $\mathcal{F}_{k-1}$ is the σ-algebra generated by $d_1, \cdots, d_{k-1}$, i.e., $\mathcal{F}_{k-1} = \sigma(d_i : 1 \leqslant i \leqslant k - 1)$. This implies that $E\{\gamma_k - \hat{\gamma}_k \mid \mathcal{F}_{k-1}\} = 0$ and hence $\{\gamma_k - \hat{\gamma}_k, \mathcal{F}_{k-1}, k \geqslant 1\}$ is a martingale difference sequence.

Since $\gamma_k \leqslant 1$, $\hat{\gamma}_k \leqslant 1$ and

$$\sum_{k=1}^{\infty} \frac{E(\gamma_k - \hat{\gamma}_k)^2}{k^2} \leqslant 4, \sum_{k=1}^{\infty} \frac{1}{k^2} < \infty$$

by Lemma A.5 we have

$$\frac{1}{k}\sum_{i=1}^{k}(\gamma_i - \hat{\gamma}_i) \to 0, \text{w. p. 1} \quad \text{as} \quad k \to \infty \tag{6.14}$$

By(6.3), define

$$\hat{\gamma}_k^0 = \sum_{j=1}^{l} I_{\{\phi_k^{\mathrm{T}} = \pi_j\}} (I_{\{C > \pi_j\theta + \chi_0\}} (1 - F(C - \pi_j\theta)) + I_{\{C \leqslant \pi_j\theta + \chi_0\}} F(C - \pi_j\theta))$$

$$= \sum_{j=1}^{l} I_{\{\phi_k^{\mathrm{T}} = \pi_j\}} \overline{F}(C - \pi_j\theta)$$

By virtue of Theorem 6.1, one can have $\hat{\gamma}_k - \hat{\gamma}_k^0 \to 0$ w. p. 1 as $k \to \infty$, which and (6.14) can lead to

$$\frac{1}{k}\sum_{i=1}^{k}(\gamma_i - \hat{\gamma}_i^0) \to 0, \text{w. p. 1} \quad \text{as} \quad k \to \infty \tag{6.15}$$

In view of(6.5), it can be seen that

$$\sum_{i=1}^{k} \hat{\gamma}_i^0 = \sum_{i=1}^{k}\sum_{j=1}^{l} I_{\{\phi_i^{\mathrm{T}} = \pi_j\}} \overline{F}(C - \pi_j\theta)$$

$$= \sum_{j=1}^{l}\left(\sum_{i=1}^{k} I_{\{\phi_i^{\mathrm{T}} = \pi_j\}}\right) \overline{F}(C - \pi_j\theta)$$

$$= \sum_{j=1}^{l} k_j \overline{F}(C - \pi_j\theta)$$

By Assumption 6.1, it follows that

$$\frac{1}{k}\sum_{i=1}^{k} \hat{\gamma}_i^0 = \sum_{j=1}^{l} \frac{k_j}{k} \overline{F}(C - \pi_j\theta) \to \sum_{j=1}^{l} \beta_j \overline{F}(C - \pi_j\theta) \quad \text{as} \quad k \to \infty$$

and the proof can be obtained from (6.15). □

Remark 6.1 By (6.7), we know that the event trigger (6.8) can achieve the communication rate in the case that θ is known.

To reflect the relationship between $\overline{\gamma}$ and C, we also write $\overline{\gamma}=\overline{\gamma}(C)$.

Lemma 6.1 $\overline{F}(z)$ from (6.3) has the following properties.

(1) $\sup_z \overline{F}(z) \leqslant \frac{1}{2}$;

(2) $\overline{F}(z)$ is monotonically increasing on $(-\infty, \mathcal{X}_0]$ and monotonically decreasing on $[\mathcal{X}_0, \infty)$.

Proof: (1) By the definition of $\overline{F}(z)$, it is known that

$$\overline{F}(z) = \frac{F(z) + (1 - F(z))}{2} - \frac{|F(z) - (1 - F(z))|}{2}$$

$$= \frac{1}{2} - \left|F(z) - \frac{1}{2}\right| \tag{6.16}$$

from which (1) is proved.

(2) Since $F(z)$ is the accumulative distribution function of d_1, $F(z)$ is monotonically increasing on $(-\infty, \infty)$. For $z_2 \leqslant z_1 \leqslant \mathcal{X}_0$, we have $F(z_2) \leqslant F(z_1)$, $F(z_1) \leqslant \frac{1}{2}$ and $F(z_2) \leqslant \frac{1}{2}$, which together with (6.16) yields

$$\overline{F}(z_1) - \overline{F}(z_2) = \left|F(z_2) - \frac{1}{2}\right| - \left|F(z_1) - \frac{1}{2}\right|$$

$$= F(z_1) - F(z_2) \geqslant 0$$

This demonstrates that $\overline{F}(z)$ is monotonically increasing on $(-\infty, \mathcal{X}_0]$. Similarly, the monotonicity on $[\mathcal{X}_0, \infty)$ can be obtained. The proof is completed. □

Proposition 6.1 Under the condition of Theorem 6.1, the communication rate $\overline{\gamma}(C)$ has the following properties.

(1) $\sup_C \overline{\gamma}(C) \leqslant \frac{1}{2}$;

(2) $\overline{\gamma}(C)$ is monotonically increasing on $(-\infty, \mathcal{X}_0+\min_{1\leqslant j\leqslant l_0}\pi_j\theta]$ and monotonically decreasing on $[\mathcal{X}_0+\max_{1\leqslant j\leqslant l_0}\pi_j\theta, \infty)$ with respect to C.

Proof: (1) By Assumption 6.1, it is known that $\sum_{j=1}^{l_0} \beta_j = 1$, which together with (1) of Lemma 6.1 implies the proof.

(2) According to (2) of Lemma 6.1, it can be seen that $\overline{F}(C - \pi_j \theta)$ is monotonically increasing on $(-\infty, \chi_0 + \pi_j \theta]$ and monotonically decreasing on $[\chi_0 + \pi_j \theta, \infty)$ with respect to C. By (6.13) and $\beta_j > 0$ for $j \in L_0$, the proof follows. □

6.4 Multi-Threshold Quantized Observations

We consider the case of multi-threshold quantized output observations. y_k is measured by a sensor of m thresholds $-\infty < C_1 < \cdots < C_m < \infty$, which can be represented by a set of m indicator functions

$$s_k = [s_{k,1}, \cdots, s_{k,m}]^{\mathrm{T}} \tag{6.17}$$

where
$$s_{k,i} = I_{\{-\infty < y_k < C_i\}}, \quad i = 1, \cdots, m$$

For every $s_{k,i}$, we use the either-or communication scheme

$$\gamma_{k,i} = \sum_{j=1}^{l} I_{\{\phi_k^{\mathrm{T}} = \pi_j\}} \left(I_{\{C \leqslant \pi_j \hat{\theta}_{k-1} + \chi_0, s_k = 1\}} + I_{\{C > \pi_j \hat{\theta}_{k-1} + \chi_0, s_k = 0\}} \right) \tag{6.18}$$

$$i = 1, \cdots, m$$

to decide whether it is transmitted to EC, as shown in Figure 6.2.

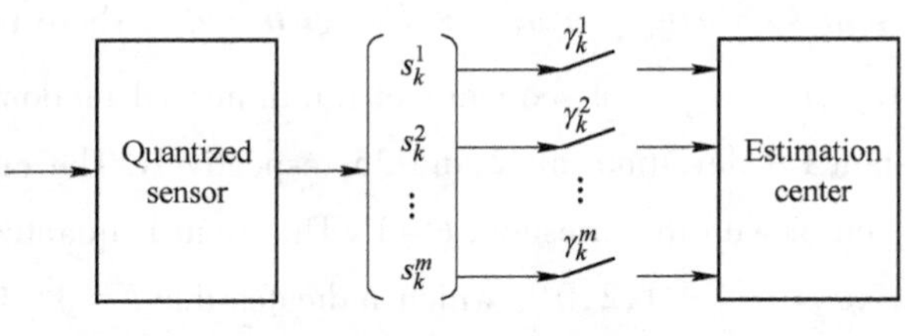

Fig. 6.2 Multi-threshold quantized observations under either-or communication

According to Theorem 6.2, it can be known that the communication rate of $\gamma_{k,i}$ is

$$\overline{\gamma}_i = \sum_{j=1}^{l_0} \beta_j \overline{F}(C_i - \pi_j \theta), \text{w. p. } 1, \quad i = 1, \cdots, m$$

For the event-triggered scheme (6.18), we can use the average communication rate of $\gamma_{k,1}, \cdots, \gamma_{k,m}$ given by

$$\bar{\gamma} = \frac{1}{m}\sum_{i=1}^{m}\bar{\gamma}_i = \frac{1}{m}\sum_{i=1}^{m}\sum_{j=1}^{l_0}\beta_j\bar{F}(C_i - \pi_j\theta)\text{, w. p. 1}$$

to indicate the ability of saving communication resources.

The available information for algorithm design is $\{\phi_k\}$, $\{\gamma_k, s_{k,i}\}$ and $\{\gamma_{k,i}\}$, $i = 1, \cdots, m$. Let

$$\alpha_{k,i} = \gamma_{k,i}s_{k,i} + \sum_{j=1}^{l} I_{\{\phi_k^{\mathrm{T}} = \pi_j\}} I_{\{C_i > \pi_j\hat{\theta}_{k-1} + x_0\}}(1 - \gamma_{k,i})$$

Then it can be seen that $[\alpha_{k,1}, \cdots, \alpha_{k,m}]^{\mathrm{T}} = s_k$ by (6.17) and (6.18). One can use $\alpha_{k,1}, \cdots, \alpha_{k,m}$ to construct m algorithms as (6.9) ~ (6.12), and obtain m estimators of θ, denoted by $\hat{\theta}_{k,1}, \cdots, \hat{\theta}_{k,m}$. Define $\nu = [\nu_1, \cdots, \nu_m]^{\mathrm{T}}$ such that $\nu_1 + \cdots + \nu_m = 1$. One can construct an estimator of θ by

$$\hat{\theta}_k = \sum_{i=1}^{m}\nu_i\hat{\theta}_{k,i}$$

which is the QCCE. Under quantized inputs, the corresponding convergence performance and the asymptotic efficiency can be given as in [41]. In order to avoid unnecessary duplication, we omit them.

6.5 Numerical Simulation

Consider an FIR system $y_k = a_1u_k + a_2u_{k-1} + d_k = \boldsymbol{\phi}_k^{\mathrm{T}}\boldsymbol{\theta} + d_k$, where the true value $\boldsymbol{\theta} = [a_1, a_2]^{\mathrm{T}} = [10, 16]^{\mathrm{T}}$ and $\{d_k\}$ is a sequence of i. i. d. normal random variables, where the mean and the standard deviation are 2 and 25 respectively. The output is measured by a binary-valued sensor with the threshold $C = 1$. The input is quantized and takes value from $\mathcal{U} = \{\mu_1, \mu_2, \mu_3\} = \{-1, 2, 0\}$, which indicates that $l = 3^2 = 9$. Let

$$\iota_{k,1} = k - \iota_{k,2} - \iota_{k,3}, \iota_{k,2} = \lceil 0.35(k - \iota_{k,3})\rceil, \iota_{k,3} = \min\{110, |\lceil \lg k\rceil|\}$$

and

$$u_k = \mu_1 I_{\{\iota_{k,1} \neq \iota_{k-1,1}\}} + \mu_2 I_{\{\iota_{k,2} \neq \iota_{k-1,2}\}} + \mu_3 I_{\{\iota_{k,3} \neq \iota_{k-1,3}\}}$$

with $\iota_{0,1} = \iota_{0,2} = \iota_{0,3} = 0$. Then it is known that $\pi_1 = [-1, -1]$, $\pi_2 = [-1, 2]$ and $\pi_3 =$

[2, -1] are persistent with $\beta_1 = 0.3$, $\beta_2 = 0.35$, $\beta_3 = 0.35$. It follows that $\beta_j = 0$ for $j = 4, \cdots, 9$, and

$$\boldsymbol{\Psi} = \begin{vmatrix} -1 & -1 \\ -1 & 2 \\ 2 & -1 \end{vmatrix}$$

is full column rank.

The either-or communication scheme (6.8) and the algorithm (6.9) ~ (6.12) is used to estimate θ with Λ being an identity matrix and $\xi_{0,j} = 1/2$ for $j \in L$. The convergence is shown by Figure 6.3.

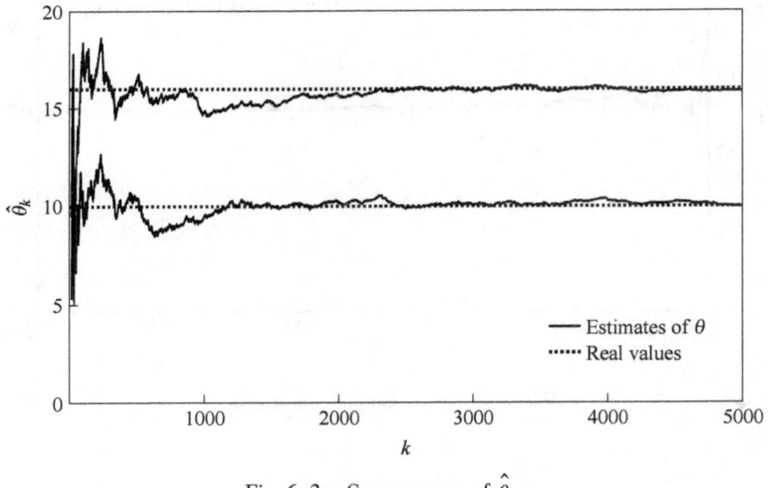

Fig. 6.3 Convergence of $\hat{\theta}_k$

Figure 6.4 provides a trajectory of the either-or communication scheme (6.8) on the time interval [200, 250], where only 12 measurements (50 ones in total) are send to EC. In Figure 6.5, it can be seen that $\frac{1}{k}\sum_{i=1}^{k}\gamma_i \to 0.2574$ as $k \to \infty$, which demonstrates the result of Theorem 6.2 with $\overline{\gamma} = \sum_{j=1}^{l_0}\beta_j \overline{F}(C - \pi_j\theta) = 0.2574$.

Note that $\mathcal{X}_0 = 2$, $\min_{1 \leqslant j \leqslant l_0}\pi_j\theta = -26$ and $\max_{1 \leqslant j \leqslant l_0}\pi_j\theta = 22$. According to Proposition 6.1, $\overline{\gamma}(C)$ is monotonically increasing on $(-\infty, 24]$ and monotonically decreasing on

[24,∞), which is displayed by Figure 6.6.

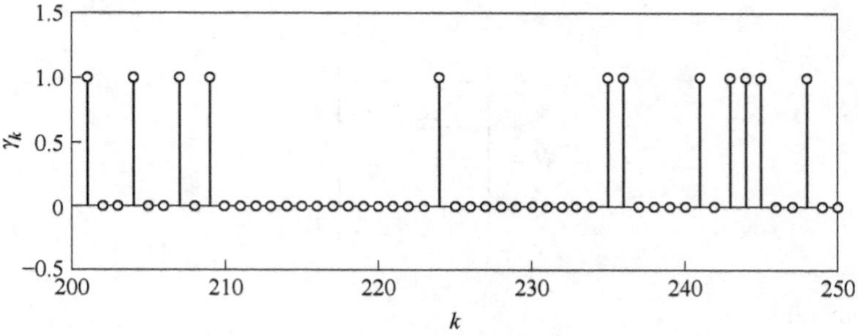

Fig. 6.4 The either-or communication scheme on [200,250]

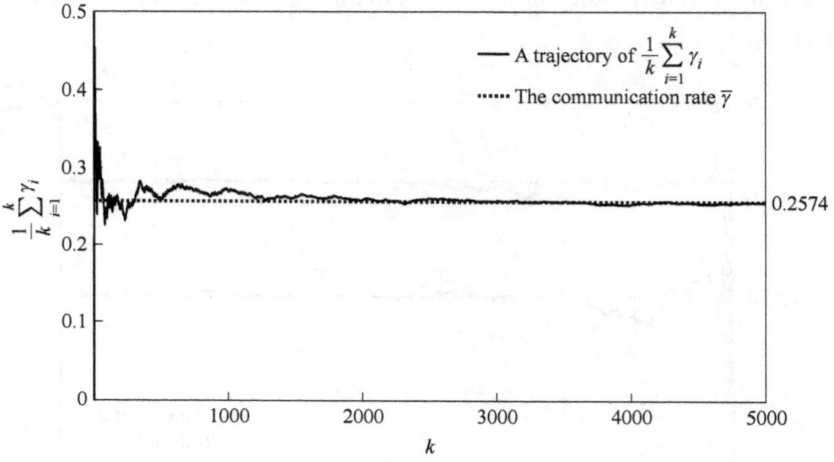

Fig. 6.5 The communication rate

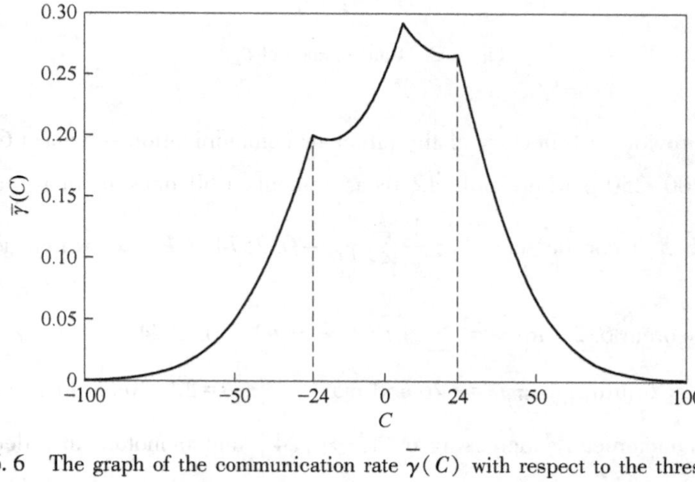

Fig. 6.6 The graph of the communication rate $\overline{\gamma}(C)$ with respect to the threshold C

6.6 Notes

This chapter addressed the quantized-input FIR system identification. To save communication resources, we introduced a so-called either-or communication scheme. An identification algorithm was designed to estimate the unknown parameter. The convergence performance of the algorithm and the communication rate was given. Then the results were extended to the case of the multi-threshold quantization.

In view of (5.14), (5.18), (6.3) and (6.4), it can be seen that

$$\overline{F}(z) = \tilde{F}(z), z \in (-\infty, \min\{\mathcal{X}_0, 0\}] \cup [\max\{\mathcal{X}_0, 0\}, +\infty) \tag{6.19}$$

which indicates that

$$\overline{F}(z) \equiv \tilde{F}(z), \text{ if } \mathcal{X}_0 = 0 \tag{6.20}$$

If $\mathcal{X}_0 < 0$, then we have $F(x) \geqslant \frac{1}{2}$, $z \in (\mathcal{X}_0, 0)$. By (5.18) and (6.4) again, it follows that $\overline{F}(z) = 1 - F(z)$, $\tilde{F}(z) = F(z)$, $z \in (\mathcal{X}_0, 0)$. Thus, one can have

$$\overline{F}(z) = 1 - F(z) \leqslant \frac{1}{2} \leqslant F(z) = \tilde{F}(z), z \in (\mathcal{X}_0, 0) \tag{6.21}$$

If $\mathcal{X}_0 > 0$, then $F(z) \leqslant \frac{1}{2}$, $\overline{F}(z) = F(z)$, $\tilde{F}(z) = 1 - F(z)$, $z \in (0, \mathcal{X}_0)$. Hence, we have

$$\overline{F}(z) = F(z) \leqslant \frac{1}{2} \leqslant 1 - F(z) = \tilde{F}(z), z \in (0, \mathcal{X}_0) \tag{6.22}$$

According to (6.19) ~ (6.22), it can be seen that we always have

$$\overline{F}(z) \leqslant \tilde{F}(z), z \in \mathbf{R}$$

which implies that

$$\sum_{j=1}^{l_0} \beta_j \overline{F}(C - \pi_j \theta) \leqslant \sum_{j=1}^{l_0} \beta_j \tilde{F}(C - \pi_j \theta)$$

Form Theorem 5. 3 and Theorem 6. 2, the above illustrates that the either-or communication scheme (6. 8) may have more strong ability in saving communication resources than the prediction-based event-triggered mechanism (5. 3). This can be understood since the statistical property of the system noise is considered sufficiently in the design of the either-or communication scheme.

7 Event-Triggered Identification of Wiener Systems with Binary-Valued Observations

There are very many different types of nonlinear systems in practice. From the perspective of control theory, system identification for nonlinear systems has been developed by focusing on specific classes of system, an important one of which is the socalled block-oriented model. Wiener system (Figure 7.1) as a model of this type was first been studied in 1958 by N. Wiener [86]. Such system is typically comprised of two blocks: a linear dynamic system followed by a nonlinear static function. Practical Wiener systems are exemplified by distillation columns [87], pH control processes [88], and biological systems [89]. Theoretically, some nonlinear systems, which are not of a Wiener structure, may be represented or approximated by a multivariate Wiener model [90]. Consequently, its study carries profound theoretical and practical significance.

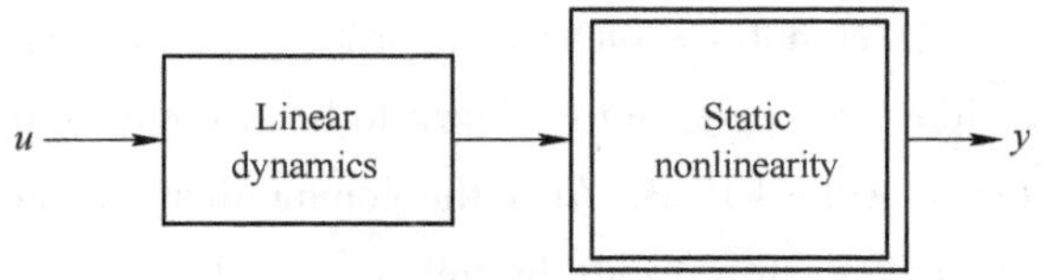

Fig. 7.1 Wiener system

Identification of Wiener systems under binary-valued/quantized observations is an interesting problem. [42] presented the first algorithm to this problem. Under scaled full-rank periodic inputs and binary-valued observations, it was showed that the identification of Wiener systems could be decomposed into a finite number of core identification problems. The concept of joint identifiability of the core problem was introduced to capture the essential conditions under which a Wiener system could be identified with binary-valued observations. A strongly convergent algorithm was constructed and proved to be asymptotically efficient for the core identification problems, achieving asymptotic optimality in its convergence rate.

However, commonly encountered inputs are not necessarily periodic. Input signals often cannot be arbitrarily selected to be periodic [91,92], and in adaptive control the control input is adjusted in real time and is usually non-periodic [82,93]. Under both quantized inputs and quantized output observations, [41] offered a constructive method to identify FIR systems, in which regressor sequences were classified into distinct pattern sets according to their values. It was shown that input-output data could be grouped, without losing any information, on the basis of both quantized output observations and input regressor patterns and used to derive an asymptotically efficient algorithm. This paper extends this idea to identify quantized-input Wiener systems under the either-or communication scheme and binary-valued output observations.

Different from the identification algorithms for linear systems, identification of Wiener systems is more complex, mainly because the internal variables between the linear and nonlinear subsystems are unmeasured, making it hard to identify the subsystems individually. In this chapter, for identifiable Wiener systems, a three-step identification algorithm is proposed. The first step aims to estimate the output of the nonlinear function by using empirical measures and organize its inputs a finite number of possible values defined as the products of basic persistent patterns and parameters of the linear dynamics. Then the second step estimates the parameters of the nonlinear function and its input values jointly. Finally, the third step estimates the parameters of the linear dynamics. Under some typical assumptions on system order, input persistent excitation, and noise distribution functions, the algorithm is shown to be strongly convergent and asymptotically efficient in terms of the CRLB. Also, the communication rate is given.

The rest of the chapter is organized into the following sections. Section 7. 1 formulates the quantized-input Wiener systems identification problem with the either-or communication scheme and binary-valued observations. System identifiability under input and output quantization is discussed in Section 7. 2. A three-step identification algorithm is introduced in Section 7. 3 based on empirical measures, persistent patterns, relations between the linear and nonlinear subsystems. Section 7. 4 establishes convergence properties of the algorithm, including strong convergence, mean-square convergence rate, and asymptotic efficiency. A numerical case study is presented in Section 7. 5 to demonstrate effectiveness of the algorithm and the convergence properties.

7. 1 Problem Formulation

Consider an SISO discrete-time Wiener system described by

$$\begin{cases} x_k = \sum_{i=1}^{n} a_i u_{k-i+1} \\ y_k = H(x_k, \eta) + d_k \end{cases} \tag{7.1}$$

where u_k , x_k and d_k are the input, the intermediate variable, and the system noise, respectively. $H(\cdot, \eta): \mathcal{D}_H \to \mathbf{R}$ is a parameterized static nonlinear function with domain $\mathcal{D}_H \subseteq \mathbf{R}$ and vector-valued parameter $\eta \in \mathbf{R}^m$. Both n and m are known. By defining the regressor $\boldsymbol{\phi}_k = [u_k, \cdots, u_{k-n+1}]^{\mathrm{T}}$ and $\boldsymbol{\theta} = [a_1, \cdots, a_n]^{\mathrm{T}}$, the linear dynamics can be expressed compactly as $x_k = \boldsymbol{\phi}_k^{\mathrm{T}} \boldsymbol{\theta}$.

The system structure is shown in Figure 7.2, in which the input u_k quantized. The output y_k is measured by a binary sensor with a finite threshold $C \in \mathbf{R}$, which can be represented by an indicator function

$$s_k = I_{\{y_k \leqslant C\}} \tag{7.2}$$

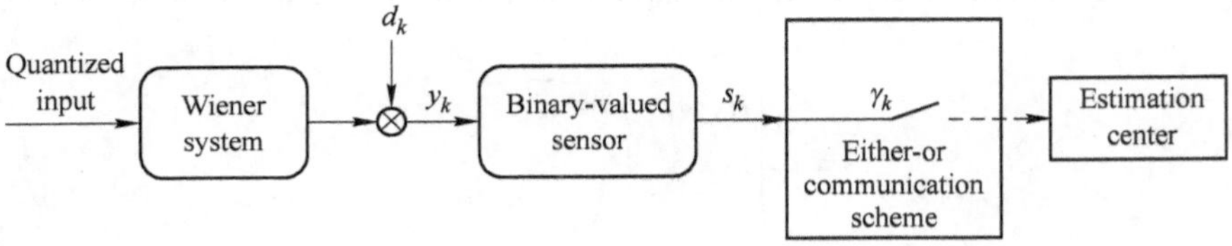

Fig. 7.2 System configuration

We use the either-or communication scheme proposed in Chapter 6 as the event-triggered mechanism, i.e. ,

$$\gamma_k = \sum_{j=1}^{l} I_{\{\phi_k^{\mathrm{T}} = \pi_j\}} \left(I_{\{C \leqslant H(\pi_j \hat{\theta}_{k-1}) + \chi_0, s_k = 1\}} + I_{\{C > H(\pi_j \hat{\theta}_{k-1}) + \chi_0, s_k = 0\}} \right) \tag{7.3}$$

where $\chi_0 = F^{-1}\left(\frac{1}{2}\right)$, and $\hat{\theta}_{k-1}$ will be given in Section 7.3.

This chapter will first discuss the issue of identifiability, then design an algorithm to identify θ and η for identifiable systems, and finally establish key convergence properties of the algorithm.

7.2 System Identifiability

System identification addresses the fundamental issue: Under what conditions, the parameters of a Wiener system can be uniquely determined from its noise-free input-output observations? For identifiable systems, algorithms can then be developed to estimate system parameters under noisy observations.

Suppose that $u = \{u_k, k = 1, 2, \cdots\}$ is an arbitrary input sequence taking quantized values in $\mathcal{U} = \{\mu_1, \cdots, \mu_r\}$. The input u generates a regressor sequence $\{\phi_k^{\mathrm{T}}\}$ that takes values in $l = r^n$ possible (row vector) patterns denoted by $\mathcal{P} = \{\pi_1, \cdots, \pi_l\}$. Denote

$$k_j = \sum_{i=1}^{k} I_{\{\phi_{n+i}^{\mathrm{T}} = \pi_j\}}, \quad j \in L = \{1, \cdots, l\}$$

Assumption 7.1 There exists $\beta_j \geqslant 0$ such that $\lim\limits_{k\to\infty} k_j/k = \beta_j$, $j \in L$. The pattern π_j is said to be *persistent* if $\beta_j > 0$. Without loss of generality, suppose that $\beta_j \neq 0$ for $j \in L_0$ and $\beta_j = 0$ for $j = L_0^-$.

It will become clear that convergence properties depend only on persistent patterns. As a result, the non-persistent patterns $\pi_{l_0+1}, \cdots, \pi_l \in L_0^-$ will not be used in designing algorithms.

Let $w_j = \pi_j \theta$. Then, the input-output mapping of the nonlinear function becomes

$$\begin{cases} H(w_1, \eta) = \bar{\alpha}_1 \\ \quad\vdots \\ H(w_{l_0}, \eta) = \bar{\alpha}_{l_0} \end{cases} \tag{7.4}$$

where the output $\bar{\alpha} = [\bar{\alpha}_1, \cdots, \bar{\alpha}_{l_0}]^{\mathrm{T}} \in \mathbf{R}^{l_0}$ can be derived from observed data and are considered to be known in study of identifiability.

Let

$$\boldsymbol{\Psi} = \begin{bmatrix} \pi_1 \\ \vdots \\ \pi_{l_0} \end{bmatrix} \tag{7.5}$$

and assume that $\boldsymbol{\Psi}$ has full column rank (then $l_0 \geqslant n$), which is a basic persistent excitation condition in identification of FIR systems with quantized inputs and outputs[41]. Hence, one can always select n patterns from $\pi_1, \cdots, \pi_{l_0}$ such that their transposes constitute a basis of $\mathbf{R}^n$. Without loss of generality, let these n patterns be $\pi_1, \cdots, \pi_n$, which will be called *the set of basic persistent patterns*.

Consequently, all other persistent patterns can be represented by the basis $\pi_j^{\mathrm{T}} = \sum_{i=1}^{n} \gamma_{j,i} \pi_i^{\mathrm{T}}$, $j = n+1, \cdots, l_0$, which implies that

$$w_j = \sum_{i=1}^{n} \gamma_{j,i} w_i, \quad j = n+1, \cdots, l_0$$

Substituting these into (7.4), we have

$$\mathbb{H}(w_1, \cdots, w_n, \eta) = \overline{\alpha} \tag{7.6}$$

with $\mathbb{H}(w_1, \cdots, w_n, \eta) = [H(\sum_{i=1}^{n} \gamma_{1,i} w_i, \eta), \cdots, H(\sum_{i=1}^{n} \gamma_{l_0,i} w_i, \eta)]^{\mathrm{T}}$.

Equality (7.6) contains l_0 equations and $n+m$ unknowns. If the solution to (7.6) exists and is unique, then η and $W := [w_1, \cdots, w_n]^{\mathrm{T}}$ can be obtained. Furthermore, θ can be derived from W since $W = \overline{\Psi}\theta$ and $\overline{\Psi}$ is full rank with $\overline{\boldsymbol{\Psi}} := [\pi_1^{\mathrm{T}}, \cdots, \pi_n^{\mathrm{T}}]^{\mathrm{T}}$. In this sense, the system (7.1) is identifiable. To ensure this identifiability in the algorithm design, we give an assumption as follows.

Assumption 7.2 There exists a compact set $\Xi \subseteq \mathbf{R}^{l_0}$ that contains the true output $\overline{\alpha} \in \Xi$ such that for any $\boldsymbol{\xi} = [\xi_1, \cdots, \xi_{l_0}]^{\mathrm{T}} \in \Xi$, the equations

$$\mathbb{H}(x_1, \cdots, x_n, \eta) = \xi \tag{7.7}$$

have a unique solution $[x_1, \cdots, x_n, \eta^{\mathrm{T}}]^{\mathrm{T}} \in \mathbf{R}^{n+m}$, denoted by $\rho(\xi)$. Moreover, $\rho(\xi)$ is bounded and continuous in Ξ.

For a given full (row) rank matrix $\boldsymbol{\Gamma} \in \mathbf{R}^{(n+m)\times l_0}$, denote its range from Ξ as $\Xi^0 = \{\zeta : \zeta = \Gamma\xi, \xi \in \Xi\} \subseteq \mathbf{R}^{n+m}$. Let

$$\mathcal{G}(x_1, \cdots, x_n, \eta) = \Gamma\mathbb{H}(x_1, \cdots, x_n, \eta) \tag{7.8}$$

Then, for any $\zeta \in \Xi^0$, there exist $\xi \in \Xi$ and $\xi^0 \in \mathbf{R}^{l_0}$ such that $\zeta = \Gamma\xi$ and $\xi = \Gamma^+ \zeta + (I - \Gamma^+ \Gamma)\xi^0$, where $^+$ represents the Moore-Penrose inverse. Under Assumption 7.2, $\mathbb{H}(\rho(\xi)) = \xi$. Hence $\mathcal{G}(\rho(\xi)) = \Gamma\,\mathbb{H}(\rho(\xi)) = \Gamma\xi = \zeta$, which implies that

$$\mathcal{G}(\rho(\Gamma^+ \zeta + (I - \Gamma^+ \Gamma)\xi^0)) = \zeta$$

Therefore, $\rho(\Gamma^{+}\zeta+(I-\Gamma^{+}\Gamma)\xi^{0})$ is a solution of the equation $\mathcal{G}(x_1,\cdots,x_n,\eta)=\zeta$. We denote it as $\tau(\zeta)=[\tau_1(\zeta),\cdots,\tau_{n+m}(\zeta)]^{\mathrm{T}}$, which means that $\tau_i(\zeta)=x_i$ for $i=1,\cdots,n$ and $[\tau_{n+1}(\zeta),\cdots,\tau_{n+m}(\zeta)]^{\mathrm{T}}=\eta$. By Assumption 7.2, $\alpha=\Gamma\overline{\alpha}\in\Xi^{0}$ and $\tau(\zeta)$ is bounded and continuous in Ξ^{0}.

In fact, (7.8) defines a linear transformation on $\mathbb{H}(x_1,\cdots,x_n,\eta)$ by the left multiplication of a full-rank matrix. This transformation has no effect on the existence of solutions to the equation (7.7). In the subsequent algorithm design, Γ will be used to improve convergence properties of the algorithm in Section 7.4.

Remark 7.1 If specific parametric models are considered on the nonlinear function, certain normalization is often needed to ensure that the parameters are independent. For example, consider the nonlinear model $y_k=b_0+b_1x_k+b_2x_k^2$ with unknown parameters $[b_0,b_1,b_2]$; and the linear dynamic system $x_k=a_1u_{k-1}$. They result in the combined system $y_k=b_0+b_1a_1u_{k-1}+b_2a_1^2u_{k-1}^2$, where there are only three independent parameters, but four unknowns. Without normalization on the scaling factor, the parameters cannot be uniquely determined from any input-output sequence. As a remedy, one may impose $b_2=1$. In this paper, we assume that such normalization has already been included in the model parameterization.

7.3 Identification Algorithm

Under Assumption 7.2, by (7.6) it is known that

$$\tau(\alpha)=[w_1,\cdots,w_n,\eta^{\mathrm{T}}]^{\mathrm{T}}=\begin{bmatrix}W\\ \\ \eta\end{bmatrix}:=\theta \tag{7.9}$$

Since this mapping $\tau(\cdot)$ is known and continuous, one may estimate α first, and then derive estimates for W and η via (7.9). With this in mind, an identification algorithm is constructed as follows, which is divided into three steps.

Identification Algorithm:

1. (Estimate α). Define $j_k^0=\sum_{j=1}^{l} jI_{\{\phi_k^{\mathrm{T}}=\pi_j\}}$

$$\eta_k=\sum_{j=1}^{l} I_{\{\phi_k^{\mathrm{T}}=\pi_j\}}I_{\{C>\pi_j\hat{\theta}_{k-1}+x_0\}}(1-\gamma_k)$$

$$\overline{S}_k^j = \begin{cases} \overline{S}_{k-1}^j, & j \neq j_k^0 \\ \left(1 - \frac{1}{k_j}\right) \overline{S}_{k-1}^j + \frac{1}{k_j}(\gamma_k s_k + \eta_k), & j = j_k^0 \end{cases} \quad (7.10)$$

$$\hat{\overline{\alpha}}_k = [C - F^{-1}(\overline{S}_k^1), \cdots, C - F^{-1}(\overline{S}_k^{l_0})]^{\mathrm{T}}$$

$$\hat{\alpha}_k = \Gamma \Pi_{\Xi}(\hat{\overline{\alpha}}_k) \quad (7.11)$$

where the initial value $\overline{S}_0^j \in (0,1)$ for $j \in L$, $\Pi_{\Xi}(z)$ is a projection from z to Ξ and can be any one in $\{\xi \in \Xi : \| \xi - z \| = \min_{\nu \in \Xi} \| \nu - z \| \}$ for $z \in \mathbf{R}^{l_0}$, and $\| \cdot \|$ is a vector norm.

2. (Estimate ϑ). Under Assumption 7.2, an estimate of θ denoted by $\hat{\theta}_k$ can be derived by

$$\hat{\theta}_k = \tau(\hat{\alpha}_k) \quad (7.12)$$

By using component-wise extraction, an estimate of W denoted by $\hat{\boldsymbol{W}}_k$ and an estimate of η denoted by $\hat{\eta}_k$ can be expressed as

$$\hat{\boldsymbol{W}}_k = [\tau_1(\hat{\alpha}_k), \cdots, \tau_n(\hat{\alpha}_k)]^{\mathrm{T}} \quad (7.13)$$

$$\hat{\boldsymbol{\eta}}_k = [\tau_{n+1}(\hat{\alpha}_k), \cdots, \tau_{n+m}(\hat{\alpha}_k)]^{\mathrm{T}} \quad (7.14)$$

3. (Estimate θ). Considering that $\hat{\boldsymbol{W}}_k = W + e_k = \overline{\Psi}\theta + e_k$ with the estimation error $e_k = \hat{W}_k - \overline{\Psi}\theta$, an estimate of θ can be constructed as

$$\hat{\theta}_k = \overline{\Psi}^{-1} \hat{W}_k \quad (7.15)$$

7.4 Convergence Properties

This section establishes key convergence properties of the identification algorithm, including strong convergence, mean-square convergence rate, and asymptotic efficiency.

7.4.1 Strong Convergence

Theorem 7.1 Consider system (7.1) with binary-valued observations (7.2). If Ψ

given by (7.5) is full column rank, and Assumptions 2.1, 7.1, and 7.2 hold, then $\hat{\theta}_k$ from (7.12) converges strongly to the true θ,

$$\hat{\theta}_k \to \theta \text{ w. p. 1} \quad \text{as} \quad k \to \infty$$

Proof: Similar to the proof of Theorem 6.1, it can be verified that $\gamma_k s_k + \eta_k = s_k$. By the strong law of large numbers and (7.10), we have $\overline{S}_k^j \to F(C - H(w_j, \eta))$, w. p. 1. as $k \to \infty$, which together with Assumption 2.1 implies that

$$C - F^{-1}(\overline{S}_k^j) \to H(w_j, \eta) \text{ w. p. 1} \quad \text{as} \quad k \to \infty, \quad j \in L_0$$

From (7.10) and (7.11), it follows that $\hat{\alpha}_k \to \alpha$ w. p. 1 as $k \to \infty$. Since $\tau(\zeta)$ is continuous in Ξ^0, $\hat{\theta}_k = \tau(\hat{\alpha}_k) \to \tau(\alpha) = \theta$ by (7.9). This completes the proof. □

Theorem 7.2 Under the conditions of Theorem 7.1, $\hat{\theta}_k$ from (7.15) converges strongly to the true θ,

$$\hat{\theta}_k \to \theta \quad \text{w. p. 1} \quad \text{as} \quad k \to \infty$$

Proof: By virtue of Theorem 7.1, we have $\hat{W}_k \to W = \overline{\Psi}\theta$, w. p. 1 as $k \to \infty$, which together with (7.15) yields that

$$\hat{\theta}_k = \overline{\Psi}^{-1}\hat{W}_k \to \overline{\Psi}^{-1}\overline{\Psi}\theta = \theta, \quad \text{w. p. 1} \quad \text{as} \quad k \to \infty$$ □

7.4.2 Asymptotic Efficiency

For convenience, denote $F^d(x) = \mathrm{d}F(x)/\mathrm{d}x$, $F_{CH}(x) = F(C - H(x, \eta))$, $F_{CH}^d(x) = F^d(C - H(x, \eta))$,

$$\Lambda = \operatorname{diag}\left[\frac{\sqrt{F_{CH}(w_j)(1 - F_{CH}(w_j))}}{F_{CH}^d(w_j)}\right]_{j=1,\cdots,l_0} \tag{7.16}$$

$\tau_j^d(\zeta) = \dfrac{\partial \tau_j(\zeta)}{\partial \zeta} = \left[\dfrac{\partial \tau_j(\zeta)}{\partial \zeta_1}, \cdots, \dfrac{\partial \tau_j(\zeta)}{\partial \zeta_{n+m}}\right]^{\mathrm{T}}$, and $J = [\tau_1^d(\alpha), \cdots, \tau_{n+m}^d(\alpha)]$ also define

$$B = \frac{\partial \mathcal{G}}{\partial \theta}, \quad D = \frac{\partial \mathbb{H}}{\partial \theta} \tag{7.17}$$

where $\mathrm{diag}[z_j]_{j=1,\cdots,\iota} = \mathrm{diag}[z_1,\cdots,z_\iota]$ is a diagonal matrix. Let $\Sigma(k;\theta)$ represent the covariance matrix of the estimation error of $\hat{\theta}_k$, i. e. ,

$$\Sigma(k;\vartheta) = E(\hat{\vartheta}_k - \vartheta)(\hat{\vartheta}_k - \vartheta)^{\mathrm{T}}, k = 1,2,\cdots$$

Lemma 7.1 If $\mathbb{H}(x_1,\cdots,x_n,\eta)$ is differentiable at θ, then the CRLB for estimating θ based on observations of $\{s_1,\cdots,s_k\}$ is

$$\Sigma_{CR}(k;\theta) = \left[\sum_{j=1}^{l} \frac{k_j(F_{CH}^d(w_j))^2 \frac{\partial H(w_j,\eta)}{\partial \theta}\left(\frac{\partial H(w_j,\eta)}{\partial \theta}\right)^{\mathrm{T}}}{F_{CH}(w_j)(1 - F_{CH}(w_j))}\right]^{-1} \tag{7.18}$$

Proof: Let z_k be some possible sample value of s_k. Since $\{d_k\}$ is i. i. d. , the likelihood function of $s_1,\cdots,s_N$ taking values $z_1,\cdots,z_k$ conditioned on θ is

$$\begin{aligned} &\ell(z_1,\cdots,z_k;\theta) \\ &= \Pr\{s_1 = z_1,\cdots,s_k = z_k;\theta\} \\ &= \prod_{i=1}^{k}[F(C - H(x_i,\eta))]^{z_i}[1 - F(C - H(x_i,\eta))]^{1-z_i} \\ &= \prod_{i=1}^{k}[F_{CH}(x_i)]^{z_i}[1 - F_{CH}(x_i)]^{1-z_i} \end{aligned}$$

Replace the particular realizations z_k by their corresponding random variables s_k, and denote the resulting quantity by $\ell = \ell(s_1,\cdots,s_k;\theta)$. Set $M_k^j = \{i:\phi_i^{\mathrm{T}} = \pi_j, 1 \leqslant i \leqslant k\}$ and $\chi_j = \frac{1}{k_j}\sum_{i\in M_k^j} s_i$. It is apparent that $E\chi_j = F_{CH}(w_j)$. Then, we have

$$\begin{aligned} \ell &= \prod_{j=1}^{l}\prod_{i\in M_k^j}[F_{CH}(w_j)]^{s_i}[1 - F_{CH}(w_j)]^{1-s_i} \\ &= \prod_{j=1}^{l}[F_{CH}(w_j)]^{k_j\chi_j}[1 - F_{CH}(w_j)]^{k_j-k_j\chi_j} \end{aligned}$$

which leads to $\lg \ell = \sum_{j=1}^{l}\{k_j\chi_j \lg F_{CH}(w_j) + (k_j - k_j\chi_j)\log[1 - F_{CH}(w_j)]\}$ and

$$\frac{\partial \log \ell}{\partial \theta}=\sum_{j=1}^{l}\left[-\frac{k_j X_j F_{CH}^d(w_j)}{F_{CH}(w_j)}\frac{\partial H(w_j,\eta)}{\partial \theta}+\frac{(k_j-k_jX_j)F_{CH}^d(w_j)}{1-F_{CH}(w_j)}\frac{\partial H(w_j,\eta)}{\partial \theta}\right]$$

Consequently, it can be verified that

$$E\frac{\partial^2 \log \ell}{\partial \theta^2}=-\sum_{j=1}^{l}\frac{k_j(F_{CH}^d(w_j))^2\frac{\partial H(w_j,\eta)}{\partial \theta}\left(\frac{\partial H(w_j,\eta)}{\partial \theta}\right)^{\mathrm{T}}}{F_{CH}(w_j)(1-F_{CH}(w_j))}$$

and (7.18) follows. □

Theorem 7.3 Under the conditions of Theorem 7.1 and Lemma 7.1, if $\tau(\zeta)$ is differentiable at α and D in (7.17) has full row rank, then the estimate $\hat{\theta}_k$ from (7.12) has the mean-square convergence rate

$$N\Sigma(k;\theta)\to(\Gamma D^{\mathrm{T}})^{-1}\Gamma Y^2\Lambda^2\Gamma^{\mathrm{T}}(D\Gamma^{\mathrm{T}})^{-1}\quad \text{as}\quad k\to\infty$$

where $Y=\mathrm{diag}[1/\sqrt{\beta_1},\cdots,1/\sqrt{\beta_{l_0}}]$ and Λ is given by (7.16).

Proof: By the mean value theorem, from (7.9) and (7.12) there exist $\tilde{\alpha}_k^1,\cdots,\tilde{\alpha}_k^{n+m}$ on the line segment $\tilde{\alpha}_k$ and α such that

$$\hat{\theta}_k-\theta=[\tau_1^d(\tilde{\alpha}_k^1),\cdots,\tau_{n+m}^d(\tilde{\alpha}_k^{n+m})]^{\mathrm{T}}(\tilde{\alpha}_k-\alpha)\tag{7.19}$$

By [17], it can be concluded that $\sqrt{k_j}(C-F^{-1}(\bar{S}_k^j)-H(w_j,\eta))\xrightarrow{d}\mathcal{N}\left(0,\frac{F_{CH}(w_j)(1-F_{CH}(w_j))}{(F_{CH}^d(w_j))^2}\right)$, which implies that

$$Y_k\begin{bmatrix}C-F^{-1}(\bar{S}_k^1)-H(w_1,\eta)\\ \vdots \\ C-F^{-1}(\bar{S}_k^{l_0})-H(w_{l_0},\eta)\end{bmatrix}\xrightarrow{d}\mathcal{N}(0,\Lambda^2)$$

as $k\to\infty$, where $Y_k=\mathrm{diag}[\sqrt{k_1},\cdots,\sqrt{k_{l_0}}]$ and $\xrightarrow{d}$ denotes convergence in distribution. In the light of Assumption 7.1, we have $Y_k/\sqrt{k}\to Y^{-1}$ as $k\to\infty$. By (7.11), one can get

$$\sqrt{k}(\tilde{\alpha}_k-\alpha)\xrightarrow{d}\mathcal{N}(0,\Gamma Y^2\Lambda^2\Gamma^{\mathrm{T}})\quad \text{as}\quad k\to\infty\tag{7.20}$$

Noting that $\tilde{\alpha}_k^j \to \alpha$ w. p. 1 as $k \to \infty$, by (7.8) and (7.19)、(7.20) it can be seen that

$$k\Sigma(k;\vartheta) \to [\tau_1^d(\alpha),\cdots,\tau_{n+m}^d(\alpha)]^{\mathrm{T}}\Gamma Y^2\Lambda^2\Gamma^{\mathrm{T}}[\tau_1^d(\alpha),\cdots,\tau_{n+m}^d(\alpha)]$$

$$= J^{\mathrm{T}}\Gamma Y^2\Lambda^2\Gamma^{\mathrm{T}}J \quad \text{as} \quad k \to \infty \tag{7.21}$$

In view of (7.8), one can have $B = D\Gamma^{\mathrm{T}}$. Since both D and Γ have full row ranks, B is full rank. Noticing that $\mathcal{G}(\tau(\alpha)) = \alpha$ and $\tau(\alpha) = \theta$, we have $BJ = JB = I$ by Theorem 12 in [94]. Therefore, $J = (D\Gamma^{\mathrm{T}})^{-1}$ and $J^{\mathrm{T}} = ((D\Gamma^{\mathrm{T}})^{-1})^{\mathrm{T}} = (\Gamma D^{\mathrm{T}})^{-1}$, which together with (7.21) implies that

$$k\Sigma(k;\vartheta) \to (\Gamma D^{\mathrm{T}})^{-1}\Gamma Y^2\Lambda^2\Gamma^{\mathrm{T}}(D\Gamma^{\mathrm{T}})^{-1} \quad \text{as} \quad k \to \infty$$

as claimed. □

Theorem 7.4 Under the conditions of Theorem 7.3, if one selects

$$\Gamma = \Gamma^* = (DY^{-2}\Lambda^{-2}D^{\mathrm{T}})^{-1}DY^{-2}\Lambda^{-2} \tag{7.22}$$

then the estimate $\hat{\theta}_k$ from (7.12) is asymptotically efficient in the sense that

$$k\Sigma(k;\vartheta) - k\Sigma_{CR}(k;\vartheta) \to 0 \quad \text{as} \quad k \to \infty$$

Proof: Since $Y, \Lambda > 0$ and D is full row rank, $DY^{-2}\Lambda^{-2}D^{\mathrm{T}} > 0$. Under the hypothesis, $\Gamma^* D^{\mathrm{T}} = (DY^{-2}\Lambda^{-2}D^{\mathrm{T}})^{-1}DY^{-2}\Lambda^{-2}D^{\mathrm{T}} = I$ by (7.22). Consequently, we have

$$(\Gamma^* D^{\mathrm{T}})^{-1}\Gamma^* Y^2\Lambda^2(\Gamma^*)^{\mathrm{T}}(D(\Gamma^*)^{\mathrm{T}})^{-1}$$

$$= \Gamma^* Y^2\Lambda^2(\Gamma^*)^{\mathrm{T}}$$

$$= (DY^{-2}\Lambda^{-2}D^{\mathrm{T}})^{-1}DY^{-2}\Lambda^{-2}D^{\mathrm{T}}((DY^{-2}\Lambda^{-2}D^{\mathrm{T}})^{-1})^{\mathrm{T}}$$

$$= (DY^{-2}\Lambda^{-2}D^{\mathrm{T}})^{-1} \tag{7.23}$$

By Lemma 7.1, it can be seen that

$$k\Sigma_{CR}(k;\theta) \to \left(\left[\frac{\partial H(w_1,\eta)}{\partial\theta},\cdots,\frac{\partial H(w_{l_0},\eta)}{\partial\theta}\right] Y^{-2}\Lambda^{-2}\times\right.$$

$$\left.\left[\frac{\partial H(w_1,\eta)}{\partial\theta},\cdots,\frac{\partial H(w_{l_0},\eta)}{\partial\theta}\right]^{\mathrm{T}}\right)^{-1}$$

$$= (DY^{-2}\Lambda^{-2}D^{\mathrm{T}})^{-1} \quad \text{as} \quad k \to \infty$$

This and (7.23) prove the theorem by virtue of Theorem 7.3. □

Theorem 7.5 Under the condition of Theorem 7.1, the communication rate from the event trigger (7.3) is given by

$$\bar{\gamma} = \sum_{j=1}^{l_0} \beta_j \tilde{F}(C - H(\pi_j \theta)), \text{ w. p. 1}$$

where $\tilde{F}(\cdot)$ is the one in (6.3).

Proof: The proof is similar to the one of Theorem 6.2 and omitted here. □

7.5 Simulation Example

Consider a Wiener system, in which the linear dynamics is a gain system and the output nonlinearity is an exponential function $H(x,\eta) = 2^x + \eta$,

$$\begin{cases} x_k = u_k \theta \\ y_k = H(x_k, \eta) + d_k = 2^{x_k} + \eta + d_k \end{cases}$$

where the true values are $\theta = 20, \eta = 30$ and $\{d_k\}$ is a sequence of i. i. d. normal random variables with zero mean and standard deviation $\sigma = 5$. The output y_k is measured by a sensor with threshold $C = 39$, and hence $s_k = I_{\{y_k \leqslant 39\}}$. The input u_k is quantized and takes values from $\mathcal{U} = \{\mu_1, \mu_2, \mu_3, \mu_4\} = \{0.1, 0.2, 3, 5\}$. Since $\theta \in \mathbf{R}$, we have $\mathcal{P} = \mathcal{U}$. Suppose that at step k, the input sequence generates patters with the following frequencies

$$k_1 = k - k_2 - k_3 - k_4, \; k_2 = \lceil 0.6(k - k_3 - k_4) \rceil$$

$$k_3 = \min\{110, |\lceil \log k \rceil|\}, \; k_4 = \lceil \sqrt{k} \rceil$$

As a result, $\beta_1 = \lim_{k\to\infty} k_1/k = 0.4, \beta_2 = 0.6, \beta_3 = \beta_4 = 0$; and $\boldsymbol{\Psi} = [1,2]^{\mathrm{T}}$, $[w_1, w_2]^{\mathrm{T}} = [2,4]^{\mathrm{T}}$, $\bar{\alpha} = [\bar{\alpha}_1, \bar{\alpha}_2]^{\mathrm{T}} = [34,46]^{\mathrm{T}}$ by (7.4) and (7.5). Let π_1 be the basic persistent pattern. Then $\pi_2 = 2\pi_1$, $w_2 = 2w_1$ and $W = w_1 = 2$.

Let $\boldsymbol{\Gamma} = I_{2\times 2}$ and $\Xi^0 = \Xi = \{[z_1, z_2]^{\mathrm{T}} \in \mathbf{R}^2 : 28 \leqslant z_1 \leqslant 40, 40 \leqslant z_2 \leqslant 52\}$. It follows that $\alpha = \bar{\alpha} \in \Xi$. For any $\xi = [\xi_1, \xi_2]^{\mathrm{T}} \in \Xi$, it can be derived that the following equations

$$\mathbb{H}(x_1,\eta)=\begin{bmatrix}H(x_1,\eta)\\ H(2x_1,\eta)\end{bmatrix}=\begin{bmatrix}\xi_1\\ \xi_2\end{bmatrix}$$

have a unique solution $\tau_1(\xi)=x_1=\log_2\left(\frac{1+\sqrt{1+4(\xi_2-\xi_1)}}{2}\right)$ and $\tau_2(\xi)=\eta=\frac{2\xi_1-1-\sqrt{1+4(\xi_2-\xi_1)}}{2}$, indicating that Assumption 7.2 holds.

The event trigger (7.3) is employed. Using (7.13) and (7.14) to compute $\hat{W}_k$ and $\hat{\eta}_k$, the convergence is shown by Figures 7.3 and 7.4. Furthermore, Figure 7.5 demonstrates the convergence of $\hat{\theta}_k$ given by (7.15).

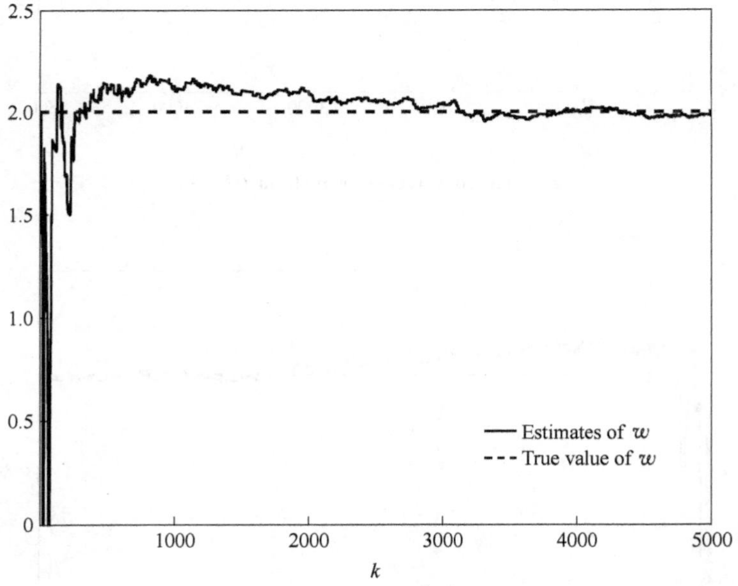

Fig. 7.3 Convergence of $\hat{W}_k$ from (7.13)

By $H(x,\eta)=2^x+\eta$, one can get

$$D=\left[\frac{\partial H(w_1,\eta)}{\partial\vartheta},\frac{\partial H(w_2,\eta)}{\partial\vartheta}\right]=\begin{bmatrix}4\ln 2 & 32\ln 2\\ 1 & 1\end{bmatrix}$$

which illustrates that D is full rank. Since $l_0=n+m=2$, and

$$\boldsymbol{\Gamma}^*=(DY^{-2}\Lambda^{-2}D^{\mathrm{T}})^{-1}DY^{-2}\Lambda^{-2}=(D^{\mathrm{T}})^{-1}$$

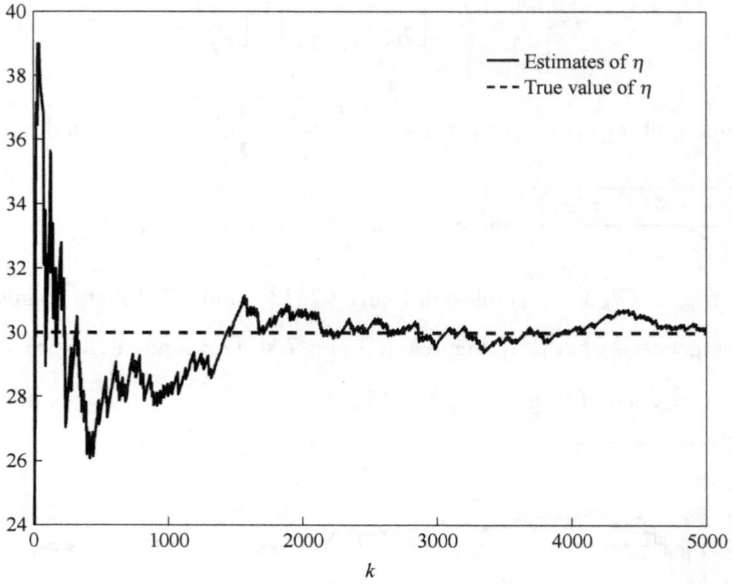

Fig. 7. 4 Convergence of $\hat{\eta}_k$ from (7. 14)

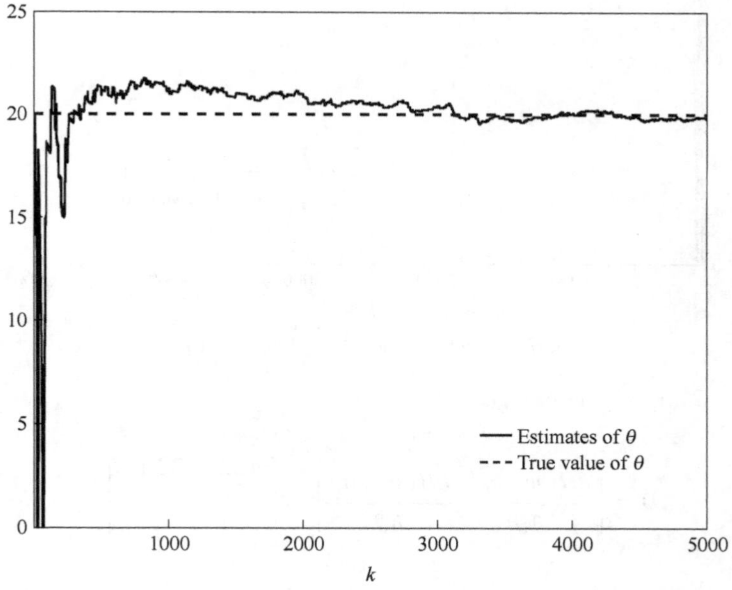

Fig. 7. 5 Convergence of $\hat{\theta}_k$ from (7. 15)

with

$$\boldsymbol{\Gamma} = \boldsymbol{\Gamma}^* = \frac{1}{28\ln 2}\begin{bmatrix} -1 & 1 \\ 32\ln 2 & -4\ln 2 \end{bmatrix}$$

Theorem 7.4 is true and hence $\hat{\theta}_k$ is asymptotically efficient, which is shown by Figure 7.6.

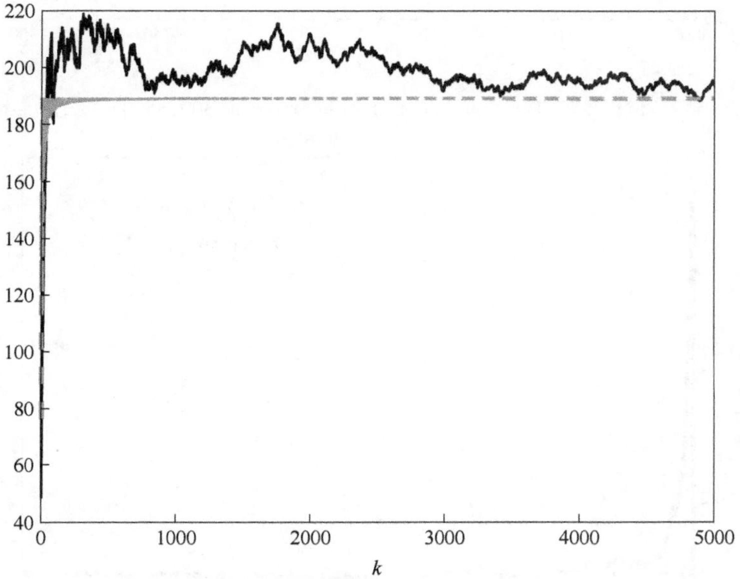

Fig. 7.6 Asymptotic efficiency of $\hat{\theta}_k$ from (7.12): The dash line is the Frobenius norm of $k\sum_{CR}(k;\theta)$ and the solid line comes from the average of 100 trajectories of the Frobenius norm of $k(\hat{\theta}_k - \theta)(\hat{\theta}_k - \theta)^{\mathrm{T}}$

Figure 7.7 shows the transmission on the time interval [300, 350], where only 8 transmissions are done. It can be inferred that $\sum_{j=1}^{l}\beta_j\tilde{F}(C - H(\pi_j\theta)) = 0.1119$, and Figure 7.8 displays that $\frac{1}{k}\sum_{i=1}^{k}\gamma_i$ converges to 0.1119, which illustrates Theorem 7.5.

7.6 Notes

Identification of Wiener systems has drawn great attention and experienced substantial advancement. Fundamental progress has been achieved in methodology development, i-

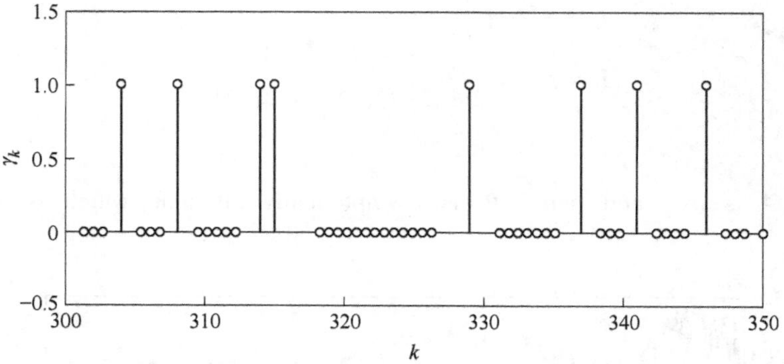

Fig. 7. 7 Transmission on the time interval[300,350]

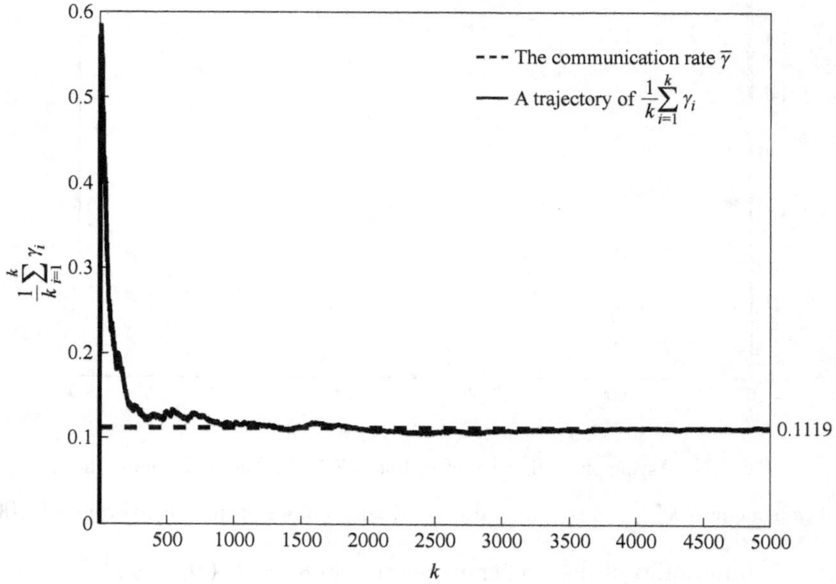

Fig. 7. 8 Communication rate

dentification algorithms, essential convergence properties, and applications [87,95~100].

This chapter studies identification of quantized-input Wiener systems under the either-or communication scheme and binary-valued output observations. After establishing identifiability conditions, a three-step algorithm is introduced to estimate unknown parameters. The algorithm is shown to be strongly convergent and achieves asymptotic efficiency in terms of the CRLB. Moreover, the communication rate is obtained. The method can potentially be extended to identify and other nonlinear systems. The next chapter will display this for Hammerstein systems with multi-threshold quantized observations.

8 Event-Triggered Identification of Hammerstein Systems with Quantized Observations

This chapter focuses on discrete-time quantized-input Hammerstein systems to investigate the identification with the prediction-based communication scheme and quantized output observations. The Hammerstein system consists of a static (memoryless) nonlinearity followed by a linear dynamic subsystem and is a class of typical and commonly encountered systems in nonlinear control applications (Figure 8. 1).

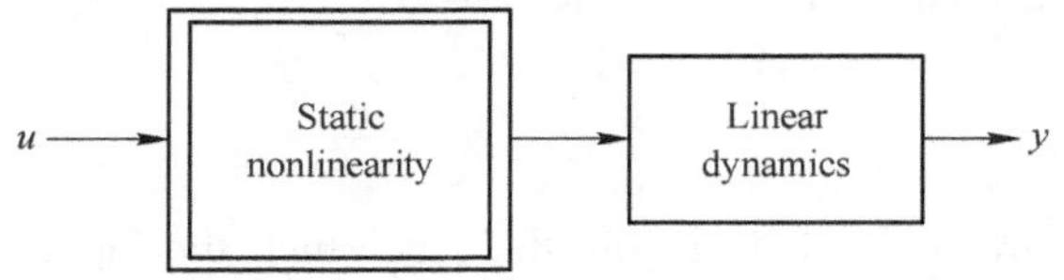

Fig. 8. 1 Hammerstein system

By taking full advantage of the communication scheme, the quantized output observations and the relationships among the inputs, an algorithm is proposed to estimate the unknown parameters based on the modified empirical measure and the quasi-convex combination technique. The output of the linear subsystem is estimated first. Then we estimate the parameters of the nonlinear function and the products of basic persistent patterns and parameters of the linear dynamics jointly. Finally, the parameters of the linear dynamics is estimated. Furthermore, the algorithm is proved to be strongly convergent and asymptotically efficient in terms of the CRLB, and the communication rate is derived.

The rest of the chapter is organized into the following sections. Section 8. 1 formulates the Hammerstein system identification problem with quantized inputs, quantized output observations and the prediction-based communication. Section 8. 2 discusses the identifiability and designs the identification algorithm for the identifiable system. Section 8. 3 establishes the properties of the algorithm, including the strong convergence, mean-square convergence rate, the asymptotic efficiency and the communication rate. Section 8. 4 simulates a numerical example to demonstrate the effectiveness of the algorithms and the main results obtained.

8.1 Problem Formulation

Consider an SISO discrete-time Hammerstein system in which the linear dynamics is an FIR system of order n_1 and the input nonlinearity is a weighted sum of n_2 functions, described by

$$\begin{cases} y_k = \sum_{i=1}^{n_1} a_i x_{k-i+1} + d_k \\ x_k = b_0 + \sum_{j=1}^{n_2} b_j h_j(u_k), b_0 = 1 \end{cases} \tag{8.1}$$

where u_k, x_k and d_k are the input, the intermediate variable, and the system noise, respectively. $h_j(\cdot)$ is a function from $\mathbf{R}$ to $\mathbf{R}$, $j \in J = \{1, 2, \cdots, n_2\}$. Both n_1 and n_2 are known.

The system structure is shown in Figure 8.2, in which the input u_k is quantized and takes a finite number of possible values from $\mathcal{U} = \{\mu_1, \cdots, \mu_r\}$. The output y_k is measured by a sensor of m thresholds $-\infty < C_1 < \cdots < C_m < \infty$. With $C_0 = -\infty$ and $C_{m+1} = \infty$, the observation sensor can be represented by

$$s_k = \sum_{i=1}^{m+1} i I_{\{y_k \in (C_{i-1}, C_i)\}} \tag{8.2}$$

whose another equivalent form is

$$\tilde{s}_k = [s_k^1, \cdots, s_k^m]^{\mathrm{T}}$$

where
$$s_k^i = I_{\{-\infty < y_k \le C_i\}}, \quad i = 1, \cdots, m \tag{8.3}$$

The triggering condition γ_k is given by

$$\gamma_k = \begin{cases} 1, s_k \neq \hat{s}_k \\ 0, s_k = \hat{s}_k \end{cases} \tag{8.4}$$

$$\hat{s}_k = \sum_{i=1}^{m+1} i I_{\{\phi_k^T \hat{\theta}_{k-1} \in (C_{i-1}, C_i]\}} \tag{8.5}$$

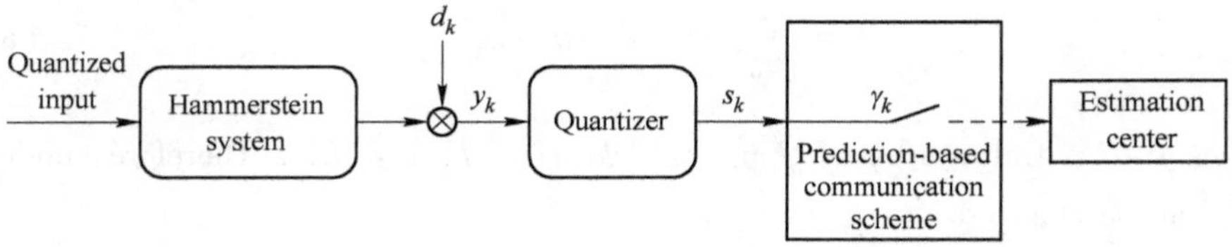

Fig. 8.2 Hammerstein system with quantized inputs and quantized output observations

Based on $\{u_k\}$ and $\{\gamma_k s_k, \gamma_k\}$, this chapter will discuss the system identifiability first, then design algorithms to estimate

$$\boldsymbol{\theta} = [a_1, \cdots, a_n]^T \quad \text{and} \quad \eta = [b_1, \cdots, b_m]^T$$

and finally establish some key convergence properties of the algorithms.

8.2 System Identifiability and Identification Algorithms

Suppose that $u = \{u_k, k \in \mathbb{N}^+\}$ is an arbitrary input sequence taking quantized values in $\mathcal{U} = \{\mu_1, \cdots, \mu_r\}$. Define $\boldsymbol{\phi}_k = [u_k, \cdots, u_{k-n_1+1}]^T$. Then the input u can generate a sequence $\{\phi_k^T\}$ that takes values in $l = r^n$ possible (row vector) patterns denoted by

$$\mathcal{P} = \{\pi_1, \cdots, \pi_l\}$$

Let $\mathbb{K}_j(k) = \{k \in \mathbb{N}^+ : \boldsymbol{\phi}_k^T = \pi_j, k \leqslant N\}$ and

$$k_j = \sum_{i=1}^{N} I_{\{\phi_{n_1+i}^T = \pi_j\}}, \quad j \in L \tag{8.6}$$

Assumption 8.1 There exists $\beta_j \geqslant 0$ such that $\lim\limits_{k\to\infty} k_j/k = \beta_j$, $j \in L$. The pattern π_j is said to be *persistent* if $\beta_j > 0$. Without loss of generality, suppose that $\beta_j \neq 0$ for $j \in L_0$ and $\beta_j = 0$ for $j = \bar{L_0}$.

Denote $h_0(\cdot) \equiv 1$, $H_{k,j}(\phi_k) = [h_j(u_k), \cdots, h_j(u_{k-n_1+1})]^T$ and $H_{k,j}(\pi_l^T) = \psi_{j,\iota}$, $j \in J^+ = \{0\} \cup J, \iota \in L$. In light of (8.1), it can be concluded that

$$y_k = \sum_{i=1}^{n_1} a_i \sum_{j=0}^{n_2} b_j h_j(u_{k-i+1}) + d_k$$

$$= \sum_{j=0}^{n_2} b_j\Big(\sum_{i=1}^{n_1} a_i h_j(u_{k-i+1})\Big) + d_k$$

$$= \sum_{j=0}^{n_2} b_j H_{k,j}^{\mathrm{T}}(\phi_k)\theta + d_k \tag{8.7}$$

Since $\phi_k^{\mathrm{T}} \in \mathcal{P}$, it always has $H_{k,j}(\phi_k) \in \{\psi_{j,\iota} : j \in J^+, \iota \in L\}$. Therefore, under π_ι (8.7) can be changed into

$$y_k = \sum_{j=0}^{n_2} b_j \psi_{j,\iota}^{\mathrm{T}} \theta + d_k \tag{8.8}$$

As seen in [41], the convergence properties mainly depend on persistent patterns and hence the non-persistent patterns will not be used in designing algorithms. There exists a known vector $\bar{\boldsymbol{\alpha}} = [\bar{\alpha}_1, \cdots, \bar{\alpha}_{l_0}]^{\mathrm{T}}$ such that

$$\sum_{j=0}^{n_2} b_j \psi_{j,\iota}^{\mathrm{T}} \theta = \bar{\alpha}_\iota, \quad \iota \in L_0 \tag{8.9}$$

Let

$$\tilde{\boldsymbol{\Psi}} = \begin{bmatrix} \pi_1 \\ \vdots \\ \pi_{l_0} \end{bmatrix} \tag{8.10}$$

and assume that $\tilde{\boldsymbol{\Psi}}$ is full column rank, which indicates that $l_0 \geqslant n_1$. As a consequence, one can select n_1 patterns from $\pi_1, \cdots, \pi_{l_0}$ such that their transpose constitute a basis of $\mathbf{R}^{n_1}$. Without loss of generality, let these n_1 ones be $\pi_1, \cdots, \pi_{n_1}$, which is called a set of *basic persistent patterns*. Noticing that $\psi_{j,\iota} \in \mathbf{R}^{l_0}$ and using $\epsilon_{j,\iota,i} \in \mathbf{R}$ to represent the coefficients of $\psi_{j,\iota}$ with respect to $\pi_1^{\mathrm{T}}, \cdots, \pi_{n_1}^{\mathrm{T}}$, we have $\psi_{j,\iota} = \sum_{i=1}^{n_1} \epsilon_{j,\iota,i} \pi_i^{\mathrm{T}}$, which gives that $\psi_{j,\iota} = \sum_{i=1}^{n_1} \epsilon_{j,\iota,i} \pi_i^{\mathrm{T}}$, $j \in J^+, \iota \in L$. This together with (8.9) yields that

$$\sum_{j=0}^{n_2} \sum_{i=1}^{n_1} b_j \epsilon_{j,\iota,i} w_i = \bar{\alpha}_\iota, \quad \iota \in L_0 \tag{8.11}$$

where $w_i = \pi_i \theta, i = 1, \cdots, n_1$. By defining

$$Y_\iota = \begin{bmatrix} \epsilon_{0,\iota,1} & \epsilon_{0,\iota,2} & \cdots & \epsilon_{0,\iota,n_1} \\ \epsilon_{1,\iota,1} & v_{1,\iota,2} & \cdots & \epsilon_{1,\iota,n_1} \\ \vdots & \vdots & & \vdots \\ \epsilon_{n_2,\iota,1} & \epsilon_{n_2,\iota,2} & \cdots & \epsilon_{n_2,\iota,n_1} \end{bmatrix} \in \mathbf{R}^{(n_2+1)\times n_1}, \quad \iota \in L_0$$

(8.11) can be simplified into

$$[b_0, b_1, \cdots, b_m] Y_\iota \begin{bmatrix} w_1 \\ \vdots \\ w_{n_1} \end{bmatrix} = \overline{\alpha}_\iota, \quad \iota \in L_0 \tag{8.12}$$

The above equality contains l_0 equations and $n_1 + n_2$ unknowns denoted by $\vartheta = [\eta^{\mathrm{T}}, W^{\mathrm{T}}]^{\mathrm{T}}$ with $W = [w_1, \cdots, w_{n_1}]^{\mathrm{T}}$. If the solution to (8.12) exists and is unique, then η and W can be obtained. Furthermore, θ can be derived by

$$\theta = \Psi^{-1} W \quad \text{with} \quad \Psi = [\pi_1^{\mathrm{T}}, \cdots, \pi_{n_1}^{\mathrm{T}}]^{\mathrm{T}} \tag{8.13}$$

Thus, the system (8.1) is identifiable in this sense. To ensure such identifiability in the algorithm design, we give an assumption as follows.

Assumption 8.2 There exists a compact set $\mathbb{W} \subseteq \mathbf{R}^{l_0}$ such that $\overline{\alpha}$ is an interior point of $\mathbb{W}$ and for any $\varpi = [\varpi_1, \cdots, \varpi_{l_0}]^{\mathrm{T}} \in \mathbb{W}$ the equations

$$\boldsymbol{\mathcal{H}}([\delta^{\mathrm{T}}, \varepsilon^{\mathrm{T}}]^{\mathrm{T}}) = \begin{bmatrix} \mathcal{H}_1([\delta^{\mathrm{T}}, \varepsilon^{\mathrm{T}}]^{\mathrm{T}}) \\ \\ \mathcal{H}_{l_0}([\delta^{\mathrm{T}}, \varepsilon^{\mathrm{T}}]^{\mathrm{T}}) \end{bmatrix} = \varpi \tag{8.14}$$

have a unique solution $[\delta^{\mathrm{T}}, \varepsilon^{\mathrm{T}}]^{\mathrm{T}} \in \mathbf{R}^{n_1+n_2}$ denoted by $\varrho(\varpi)$, where $\boldsymbol{\delta} = [\delta_1, \cdots, \delta_{n_2}]^{\mathrm{T}} \in \mathbf{R}^{n_2}$, $\boldsymbol{\varepsilon} = [\varepsilon_1, \cdots, \varepsilon_{n_1}]^{\mathrm{T}} \in \mathbf{R}^{n_1}$ and

$$\mathcal{H}_\iota([\delta^{\mathrm{T}}, \varepsilon^{\mathrm{T}}]^{\mathrm{T}}) = [1, \delta^{\mathrm{T}}] Y_\iota \varepsilon, \quad \iota \in L_0 \tag{8.15}$$

Moreover, $\varrho(\varpi)$ is bounded and continuous in $\mathbb{W}$.

For a given full (row) rank matrix $\mathcal{L} \in \mathbf{R}^{(n_1+n_2)\times l_0}$, denote its range from $\mathbb{W}$ as $\mathbb{W}_0 = \{\varsigma : \varsigma = \mathcal{L}\varpi, \varpi \in \mathbb{W}\} \subseteq \mathbf{R}^{n_1+n_2}$. Let

$$\mathcal{G}([\delta^{\mathrm{T}},\varepsilon^{\mathrm{T}}]^{\mathrm{T}}) = \mathcal{L}\mathcal{H}([\delta^{\mathrm{T}},\varepsilon^{\mathrm{T}}]^{\mathrm{T}})$$

Then, for any $\varsigma \in \mathbb{W}_0$, there exist $\varpi \in \mathbb{W}$ and $\varpi^0 \in \mathbf{R}^{l_0}$ such that $\varsigma = \mathcal{L}\varpi$ and $\varpi = \mathcal{L}^+\varsigma + (I - \mathcal{L}^+\mathcal{L})\varpi^0$, where $^+$ represents the Moore-Penrose inverse and I is the identity matrix of suitable dimension. Under Assumption 8.2, we have $\mathcal{H}(\varrho(\varpi)) = \varpi$, which implies that $\mathcal{G}(\varrho(\varpi)) = \mathcal{L}\mathcal{H}(\varrho(\varpi)) = \mathcal{L}\varpi = \varsigma$, and

$$\mathcal{G}(\varrho(\mathcal{L}^+\varsigma + (I - \mathcal{L}^+\mathcal{L})\varpi^0)) = \varsigma$$

Therefore, $\varrho(\mathcal{L}^+\varsigma + (I - \mathcal{L}^+\mathcal{L})\varpi^0)$ is a solution of the equation $\mathcal{G}([\delta^{\mathrm{T}},\varepsilon^{\mathrm{T}}]^{\mathrm{T}}) = \varsigma$. We denote it as $\tau(\varsigma) = [\tau_1(\varsigma),\cdots,\tau_{n_1+n_2}(\varsigma)]^{\mathrm{T}}$, which means that $\tau_i(\varsigma) = \delta_i$ for $i = 1,\cdots,n_2$ and $\tau_i(\varsigma) = \varepsilon_i$ for $i = n_2+1,\cdots,n_2+n_1$. According to Assumption 8.2, $\alpha = \mathcal{L}\overline{\alpha} \in \mathbb{W}_0$ and $\tau(\varsigma)$ is bounded and continuous in $\mathbb{W}_0$. In addition, by (8.12) it can be seen that $\varrho(\overline{\alpha}) = \vartheta$ and

$$\tau(\alpha) = \vartheta \tag{8.16}$$

For the identifiable system, the following three-step algorithm is introduced to estimate the unknown parameters. The first step aims to estimate α and yields $\hat{\alpha}(k)$, based on which the seconde step estimates $\vartheta = [\eta^{\mathrm{T}}, W^{\mathrm{T}}]^{\mathrm{T}}$. Finally, the third step does θ using the estimate of W.

Identification Algorithm:

(1) (Estimate α). Define $j_k^0 = \sum_{j=1}^{l} jI_{\{\phi_k^{\mathrm{T}} = \pi_j\}}$, $\eta_k = \gamma_k s_k + (1-\gamma_k)\hat{s}_k$ and

$$\eta_k^i = \begin{cases} i, & \eta_k \leqslant i \\ 0, & \text{otherwise} \end{cases} \tag{8.17}$$

$$\overline{S}_{j,i}(k) = \begin{cases} \overline{S}_{j,i}(k-1), & j \neq j_k^0 \\ \left(1 - \dfrac{1}{k_j}\right)\overline{S}_{j,i}(k-1) + \dfrac{1}{k_j}\eta_k^i, & j = j_k^0 \end{cases} \quad i = 1,\cdots,m \tag{8.18}$$

Let $\nu_j = [\nu_{j,1},\cdots,\nu_{j,m}]^{\mathrm{T}} \in \mathbf{R}^m$ such that $\nu_{j,1} + \cdots + \nu_{j,m} = 1, j \in L_0$. An estimate of α denoted by $\hat{\alpha}(k)$ is constructed as follows

$$\hat{\overline{\alpha}}_{j,i}(k) = C_i - F^{-1}(S_{j,i}(k)), \quad i = 1,\cdots,m \tag{8.19}$$

$$\hat{\bar{\alpha}}_j(k) = \sum_{i=1}^{m} \nu_{j,i}\, \hat{\bar{\alpha}}_{j,i}(k), \quad j \in L_0 \tag{8.20}$$

$$\hat{\bar{\alpha}}(k) = [\hat{\bar{\alpha}}_1(k), \cdots, \hat{\bar{\alpha}}_{l_0}(k)]^{\mathrm{T}} \tag{8.21}$$

$$\hat{\alpha}(k) = \mathcal{L}\prod\nolimits_{\mathbb{W}}(\hat{\bar{\alpha}}(k)) \tag{8.22}$$

where $\prod_{\mathbb{W}}(z)$ is a projection from z to $\mathbb{W}$ and can be any one in $\{\omega \in \mathbb{W}: \|\omega - z\| = \min_{\nu \in \mathbb{W}} \|\nu - z\|\}$ for $z \in \mathbf{R}^{l_0}$.

(2) (Estimate ϑ). Under Assumption 8.2, an estimate of ϑ denoted by $\hat{\vartheta}(k)$ can be derived by

$$\hat{\vartheta}(k) = \tau(\hat{\alpha}(k)) \tag{8.23}$$

By using component-wise extraction, an estimate of η denoted by $\hat{\eta}(k)$ and an estimate of W denoted by $\hat{W}(k)$ and can be expressed as

$$\hat{\eta}(k) = [\tau_1(\hat{\alpha}(k)), \cdots, \tau_{n_2}(\hat{\alpha}(k))]^{\mathrm{T}} \tag{8.24}$$

$$\hat{W}(k) = [\tau_{n_2+1}(\hat{\alpha}(k)), \cdots, \tau_{n_2+n_1}(\hat{\alpha}(k))]^{\mathrm{T}} \tag{8.25}$$

(3) (Estimate θ). Considering (8.13) and $\hat{W}(k) = W + e(k) = \Psi\theta + e(k)$ with the estimation error $e(k) = \hat{W}(k) - \Psi\theta$, an estimate of θ denoted by $\hat{\theta}(k)$ can be given as

$$\hat{\theta}(k) = \Psi^{-1}\hat{W}(k) \tag{8.26}$$

8.3 Convergence Properties

This section establishes the convergence properties of the identification algorithm, including the strong convergence, the mean-square convergence rate and the asymptotical efficiency. Let $\Sigma(k;\vartheta)$ represent the covariance matrix of the estimation error of $\hat{\vartheta}(k)$, i.e.,

$$\Sigma(k;\vartheta) = E(\hat{\vartheta}(k) - \vartheta)(\hat{\vartheta}(k) - \vartheta)^{\mathrm{T}}$$

Theorem 8.1 Consider system (8.1) with quantized inputs $u \in \mathcal{U}$ and quantized observations (8.3). If $\tilde{\Psi}$ given by (8.10) is full column rank, and Assumptions 2.1, 8.1

and 8.2 hold, then $\hat{\vartheta}(k)$ from (8.23) and $\hat{\theta}(k)$ from (8.26) converge strongly to the true ϑ and θ,

$$\hat{\vartheta}(k) \to \vartheta,\ \hat{\theta}(k) \to \theta \quad \text{w. p. 1} \quad \text{as} \quad k \to \infty$$

Proof: In view of (8.17), (8.2)~(8.5), it can be seen that $[\eta_k^1, \cdots, \eta_k^m]^{\mathrm{T}} = \tilde{s}_k$. Noticing that $\boldsymbol{\phi}_k^{\mathrm{T}} = \pi_\iota$ for $k \in \mathbb{K}_\iota(k)$ and $\mathbb{K}_\iota(k)$ has k_ι elements, by the law of large numbers one can have

$$S_{\iota,i}(k) \to F\Big(C_i - \sum_{j=0}^{n_2} b_j \psi_{j,\iota}^{\mathrm{T}} \theta\Big) = F(C_i - \bar{\alpha}_\iota)$$

$$\text{as} \quad k \to \infty,\ i = 1, \cdots, m,\ \iota \in L_0$$

from (8.8), (8.9) and (8.18). In light of (8.19) and (8.20), it can be seen that $\hat{\bar{\alpha}}_{\iota,i}(k) \to \bar{\alpha}_\iota$ and then $\hat{\bar{\alpha}}_\iota(k) \to \bar{\alpha}_\iota$ as $k \to \infty$ since $\nu_{\iota,1} + \cdots + \nu_{\iota,m} = 1, \iota \in L_0$, Consequently, by (8.21) we have

$$\hat{\bar{\alpha}}(k) \to [\bar{\alpha}_1, \cdots, \bar{\alpha}_{l_0}]^{\mathrm{T}} = \bar{\alpha}$$

which together with (8.22) gives that $\hat{\alpha}(k) \to \alpha$ as $k \to \infty$. By (8.16) and (8.23), it can be concluded that $\hat{\vartheta}(k) \to \tau(\alpha) = \vartheta$ as $k \to \infty$. Hence, by (8.25) one can know that $\hat{W}(k) \to W$, which implies that $\hat{\theta}(k) \to \theta$ as $k \to \infty$ by (8.13) and (8.26). This completes the proof. □

Theorem 8.2 Under the conditions of Theorem 8.1, if $\tau(\varsigma)$ is differentiable at α, then the estimate $\hat{\vartheta}(k)$ from (8.23) has the mean-square convergence rate in the sense that

$$k\Sigma(k;\vartheta) \to \Gamma^{\mathrm{T}} \mathcal{L} \Lambda^2 \Delta^2 \mathcal{L}^{\mathrm{T}} \Gamma \quad \text{as} \quad k \to \infty$$

where $\Gamma = \dfrac{\partial \tau(\alpha)}{\partial \alpha}$, $\Lambda = \mathrm{diag}[1/\sqrt{\beta_1}, \cdots, 1/\sqrt{\beta_{l_0}}]$ with $\beta_1, \cdots, \beta_{l_0}$ being the ones in Assumption 8.1, and $\Delta^2 = \mathrm{diag}[\nu_\iota^T U(\mathcal{H}_\iota(\vartheta)) P(\mathcal{H}_\iota(\vartheta)) U(\mathcal{H}_\iota(\vartheta)) \nu_\iota]_{\iota=1,\cdots,l_0}$ with $U(\cdot)$ and $P(\cdot)$ being given by (2.3) and (2.4).

Proof: By the mean-value theorem, there exist $\tilde{\alpha}^1(k), \cdots, \tilde{\alpha}^{n_1+n_2}(k)$ on the line segment $\tilde{\alpha}(k)$ and α such that

$$\hat{\vartheta}(k) - \vartheta = [\tau_1^d(\tilde{\alpha}^1(k)), \cdots, \tau_{n_1+n_2}^d(\tilde{\alpha}^{n_1+n_2}(k))]^{\mathrm{T}}(\hat{\alpha}(k) - \alpha) \qquad (8.27)$$

from (8.16) and (8.23). From (8.12), (8.15) and [17], we know that $\bar{\alpha}_\iota = \mathcal{H}_\iota(\vartheta)$, $\bar{\alpha} = [\mathcal{H}_1(\vartheta), \cdots, \mathcal{H}_{l_0}(\vartheta)]^{\mathrm{T}}$ and

$$\sqrt{k_\iota}(\hat{\bar{\alpha}}_\iota(k) - \bar{\alpha}_\iota) \xrightarrow{d} \mathcal{N}(0, \nu_\iota^{\mathrm{T}} U(\mathcal{H}_\iota(\vartheta)) V(\mathcal{H}_\iota(\vartheta)) U(\mathcal{H}_\iota(\vartheta)) \nu_\iota), \quad \iota \in L_0$$

which implies that

$$\mathrm{diag}[\sqrt{k_1}, \cdots, \sqrt{k_{l_0}}](\hat{\bar{\alpha}}(k) - \bar{\alpha}) \xrightarrow{d} \mathcal{N}(0, \Delta^2) \quad \text{as} \quad k \to \infty \qquad (8.28)$$

Noticing that $\mathrm{diag}[\sqrt{k_1}, \cdots, \sqrt{k_{l_0}}]/\sqrt{k} = \Lambda^{-1}$ as $k \to \infty$ according to Assumption 8.1 and both Λ and Δ^2 are diagonal matrixes, by (8.22) and (8.28) we have

$$\sqrt{k}(\hat{\alpha}(k) - \alpha) \xrightarrow{d} \mathcal{N}(0, \mathcal{L}\Lambda^2\Delta^2\mathcal{L}^{\mathrm{T}}) \qquad (8.29)$$

From Theorem 8.1, it can be seen that $\tilde{\alpha}^j(k) \to \alpha, j = 1, \cdots, n_1 + n_2$. In view of (8.27) and (8.29), one can have

$$k\Sigma(k;\vartheta) \to [\tau_1^d(\alpha), \cdots, \tau_{n_1+n_2}^d(\alpha)]^{\mathrm{T}} \Delta\Lambda^2\Delta^2\mathcal{L}^{\mathrm{T}} [\tau_1^d(\alpha), \cdots, \tau_{n_1+n_2}^d(\alpha)]$$

$$= \Gamma^{\mathrm{T}}\mathcal{L}\Lambda^2\Delta^2\mathcal{L}^{\mathrm{T}}\Gamma \quad \text{as} \quad k \to \infty$$

as claimed. □

Theorem 8.3 The CRLB for estimating ϑ based on observations of $\{\tilde{s}_1, \cdots, \tilde{s}_k\}$ is

$$\Sigma_{CR}(k;\vartheta) = \left[\sum_{j=1}^{l}\sum_{i=1}^{m+1} k_j \frac{(\zeta_i^d(\mathcal{H}_j(\vartheta)))^2}{\zeta_i(\mathcal{H}_j(\vartheta))} \frac{\partial \mathcal{H}_j(\vartheta)}{\partial \vartheta}\left(\frac{\partial \mathcal{H}_j(\vartheta)}{\partial \vartheta}\right)^{\mathrm{T}}\right]^{-1}$$

where k_j is defined by (8.6), $\zeta_i(\cdot) = F_i(\cdot) - F_{i-1}(\cdot)$ and $\zeta_i^d(\cdot) = \frac{\mathrm{d}}{\mathrm{d}x}\zeta_i(\cdot), i = 1, \cdots, m, j \in L$.

Proof: Augment $\tilde{s}_k$ to $\bar{s}_k = [s_k^{\mathrm{T}}, 1]^{\mathrm{T}} \in \mathbf{R}^{m+1}$, where the added element represents $1 = \Pr\{-\infty < y_k < \infty\}$. Let z_k be some possible sample value of $\bar{s}_k$. Noticing that z_k always takes the form of $[0, \cdots, 0, 1, 1, \cdots, 1]^{\mathrm{T}}$, we have

$$\Pr\{\bar{s}_k = z_k; \vartheta\}$$

$$= \Pr\{ C_{i_0(k)-1} < y_k \leqslant C_{i_0(k)} \}$$

$$= F_{i_0(k)}\Big(\sum_{j=0}^{n_2} b_j H_{k,j}^{\mathrm{T}}(\phi_k)\theta \Big) - F_{i_0(k)-1}\Big(\sum_{j=0}^{n_2} b_j H_{k,j}^{\mathrm{T}}(\phi_k)\theta \Big)$$

$$= \zeta_{i_0(k)}\Big(\sum_{j=0}^{n_2} b_j H_{k,j}^{\mathrm{T}}(\phi_k)\theta \Big)$$

where $i_0(k)$ means the index of the first 1 in z_k. Since $\{d_k\}$ is i. i. d., the likelihood function of $\bar{s}_1, \cdots, \bar{s}_N$ taking values $z_1, \cdots, z_N$, conditioned on ϑ, is

$$\ell(z_1, \cdots, z_k; \vartheta) = \Pr\{ \bar{s}_1 = z_1, \cdots, \bar{s}_k = z_k; \vartheta \}$$

$$= \prod_{i=1}^{k} \zeta_{i_0(i)}\Big(\sum_{j=0}^{n_2} b_j H_{i,j}^{\mathrm{T}}(\phi_i)\theta \Big)$$

Replace the particular realizations z_k by their corresponding random elements $\bar{s}_k$, and denote the resulting quantity by $\ell = \ell(\bar{s}_1, \cdots, \bar{s}_k; \vartheta)$. With $\chi_{j,i} = \frac{1}{k_j} \sum_{k \in \mathbb{K}_j(k)} s_k^i$, $l = 1, \cdots, m$, $j \in L$, by (8.8), (8.12) and (8.15) we have

$$\ell = \prod_{j=1}^{l} \prod_{i \in \mathbb{K}_j(k)} \zeta_{i_0(i)}(H_j(\vartheta)) = \prod_{j=1}^{l} \prod_{i=1}^{m+1} (\zeta_i(\mathcal{H}_j(\vartheta)))^{k_j \chi_{j,i}}$$

Then it follows that $\frac{\partial}{\partial \vartheta} \log \ell = \sum_{j=1}^{l} \sum_{i=1}^{m+1} k_j \chi_{j,i} \frac{\zeta_i^d(\mathcal{H}_j(\vartheta))}{\zeta_i(\mathcal{H}_j(\vartheta))} \frac{\partial \mathcal{H}_j(\vartheta)}{\partial \vartheta}$ and

$$\frac{\partial^2}{\partial \vartheta^2} \log \ell = \sum_{j=1}^{l} \sum_{i=1}^{m+1} k_j \chi_{j,i} \frac{1}{\zeta_i(\mathcal{H}_j(\vartheta))} \frac{\partial}{\partial \vartheta}\left(\zeta_i^d(\mathcal{H}_j(\vartheta)) \frac{\partial \mathcal{H}_j(\vartheta)}{\partial \vartheta} \right) -$$

$$\sum_{j=1}^{l} \sum_{i=1}^{m+1} k_j \chi_{j,i} \left(\frac{\zeta_i^d(\mathcal{H}_j(\vartheta))}{\zeta_i(\mathcal{H}_j(\vartheta))} \right)^2 \frac{\partial \mathcal{H}_j(\vartheta)}{\partial \vartheta} \left(\frac{\partial \mathcal{H}_j(\vartheta)}{\partial \vartheta} \right)^{\mathrm{T}}$$

$$:= \boldsymbol{I}_1 - \boldsymbol{I}_2 \tag{8.30}$$

Since $\sum_{i=1}^{m+1} \zeta_i = 1$ and $E_{\chi_{j,i}} = \zeta_i(\mathcal{H}_j(\vartheta))$, one can obtain that

$$\boldsymbol{EI}_1 = \sum_{j=1}^{l} \sum_{i=1}^{m+1} k_j \frac{\partial}{\partial \vartheta}\left(\zeta_i^d(\mathcal{H}_j(\vartheta)) \frac{\partial \mathcal{H}_j(\vartheta)}{\partial \vartheta} \right)$$

$$= \sum_{j=1}^{l} k_j \frac{\partial^2}{\partial \vartheta^2}\Big(\sum_{i=1}^{m+1} \zeta_i(\mathcal{H}_j(\vartheta))\Big) = 0 \tag{8.31}$$

and

$$\boldsymbol{EI}_2 = \sum_{j=1}^{l}\sum_{i=1}^{m+1} k_j \frac{(\zeta_i^d(\mathcal{H}_j(\vartheta)))^2}{\zeta_i(\mathcal{H}_j(\vartheta))}\frac{\partial \mathcal{H}_j(\vartheta)}{\partial \vartheta}\left(\frac{\partial \mathcal{H}_j(\vartheta)}{\partial \vartheta}\right)^{\mathrm{T}} \tag{8.32}$$

On account of $\Sigma_{CR}(k;\vartheta) = \left(-E\frac{\partial^2}{\partial\vartheta^2}\lg\ell\right)^{-1}$, by (8.30) ~ (8.32) the theorem is proved. □

Theorem 8.4 Under the conditions of Theorem 8.1, if

$$\Xi = \left[\frac{\partial \mathcal{H}_1(\vartheta)}{\partial \vartheta}, \cdots, \frac{\partial \mathcal{H}_{l_0}(\vartheta)}{\partial \vartheta}\right]^{\mathrm{T}}$$

is full rank, $P_\iota(k;\vartheta)$ is positive definite, $\nu_j = \lambda_j^* = \dfrac{P_j^{-1}(k;\vartheta)\boldsymbol{1}}{1^{\mathrm{T}}P_j^{-1}(k;\vartheta)1}$ for $j \in L_0$ and one selects

$$\mathcal{L} = \mathcal{L}^* = (\Xi^{\mathrm{T}}\Lambda^{-2}(\Delta^*)^{-2}\Xi)^{-1}\Xi^{\mathrm{T}}\Lambda^{-2}(\Delta^*)^{-2}$$

then the estimate $\hat{\vartheta}(k)$ from (8.23) is asymptotically efficient in the sense that

$$k\Sigma(k;\vartheta) - k\Sigma_{CR}(k;\vartheta) \to 0 \quad \text{as} \quad k \to \infty$$

where $P_j(k;\vartheta) = E(e_j(k;\vartheta))(e_j(k;\vartheta))^{\mathrm{T}}$ with $e_j(k;\vartheta) = [\hat{\bar{\alpha}}_{j,1}(k), \cdots, \hat{\bar{\alpha}}_{j,m}(k)]^{\mathrm{T}} - H_j(\vartheta)\boldsymbol{1}$ for $j \in L_0$, and

$$\Delta^* = \left(\mathrm{diag}\left[\sqrt{\sum_{i=1}^{m+1}\frac{(\zeta_i^d(\mathcal{H}_j(\vartheta)))^2}{\zeta_i(\mathcal{H}_j(\vartheta))}}\right]_{j=1,\cdots,l_0}\right)^{-1}$$

Proof: Under hypothesis, $\mathcal{L}^*$ is well defined. From [17], we know that the optimal QCCE is used in (8.20) and

$$k\Sigma(k;\vartheta) \to \Gamma^{\mathrm{T}}\mathcal{L}^*\Lambda^2(\Delta^*)^2(\mathcal{L}^*)^{\mathrm{T}}\Gamma \quad \text{as} \quad k \to \infty \tag{8.33}$$

in virtue of Theorem 8.2. From (8.14) and (8.16), we have $\mathcal{L}\mathcal{H}(\tau(\alpha)) = \alpha$, which indicates that $\mathcal{L}\Xi\Gamma = I$. Together with (8.33), it follows that

$$k\Sigma(k;\vartheta) \to (\Xi^{\mathrm{T}}(\mathcal{L}^*)^{\mathrm{T}})^{-1}\mathcal{L}^*\Lambda^2(\Delta^*)^2(\mathcal{L}^*)^{\mathrm{T}}(\mathcal{L}^*\Xi)^{-1} \quad \text{as} \quad k \to \infty \tag{8.34}$$

Moreover, one can verify that $\mathcal{L}^*\Xi = \Xi^{\mathrm{T}}(\mathcal{L}^*)^{\mathrm{T}} = I, \Lambda^2\Delta^2(\mathcal{L}^*)^{\mathrm{T}} = \Xi(\Xi^{\mathrm{T}}\Lambda^{-2}(\Delta^*)^{-2}\Xi)^{-1}$, and

$$\begin{aligned}&(\Xi^{\mathrm{T}}(\mathcal{L}^*)^{\mathrm{T}})^{-1}\mathcal{L}^*\Lambda^2(\Delta^*)^2(\mathcal{L}^*)^{\mathrm{T}}(\mathcal{L}^*\Xi)^{-1}\\ &= \mathcal{L}^*\Lambda^2(\Delta^*)^2(\mathcal{L}^*)^{\mathrm{T}} = (\Xi^{\mathrm{T}}A^{-2}(\Delta^*)^{-2}\Xi)^{-1}\end{aligned}$$

From this and (8.34), we have

$$k\Sigma(k;\vartheta) \to (\Xi^{\mathrm{T}}\Lambda^{-2}(\Delta^*)^{-2}\Xi)^{-1} \quad \text{as} \quad k \to \infty \tag{8.35}$$

On the other hand, by Theorem 8.3 and Assumption 8.1 it can be derived that

$$k\Sigma_{CR}(k;\vartheta) \to (\Xi^{\mathrm{T}}\Lambda^{-2}(\Delta^*)^{-2}\Xi)^{-1} \quad \text{as} \quad k \to \infty$$

which together with (8.35) gives the proof. □

Remark 8.1 In this paper, one key step is the empirical-measure-based estimation (8.17) ~ (8.22), whose convergence analysis is based on Assumption 2.1where the system noise is i.i.d. In fact, such estimation algorithm still can work for the correlated noise[101]. Then it is possible to handle the identification problem for the case of colored noises by employing the method in [101]. The main difficulties are to deal with the correlation caused by the correlated noise and derive the explicit representation of the CRLB.

Theorem 8.5 Under the condition of Theorem 8.1, $\overline{\gamma}_k$ from (8.4) converges to

$$\overline{\gamma} = \sum_{\iota=1}^{l_0}\beta_j\sum_{i=1}^{m+1}\left(I_{\{\sum_{j=0}^{n_2} b_j\psi_{j,\iota}^{\mathrm{T}}\theta \notin (C_{i-1},C_i]\}} \times (F(C_i - \pi_j\theta) - F(C_{i-1} - \pi_j\theta))\right), \text{w. p. } 1$$

Proof: The proof is similar to the one of Theorem 5.3 and omitted here. □

8.4 Simulation

Consider a Hammerstein system ($n_1 = n_2 = 1$), in which the linear dynamics is a gain system and the input nonlinearity is formed by an power function and an exponential function

$$\begin{cases} y_k = \theta x_k + d_k \\ x_k = b_0 + \eta \cdot \dfrac{u_k^2 - 2^{u_k}}{9}, \quad b_0 = 1 \end{cases}$$

where the true values are $\theta = 11, \eta = 9$ and $\{d_k\}$ is a sequence of i. i. d. normal random variables with zero mean and standard deviation $\sigma = 4$. The output y_k is measured by a sensor with the single threshold $C = 17$, and hence $m = 1$ and $s_k = I_{\{y_k \leqslant C\}}$. The event-trigger (8.4) is used to decide wether s_k is send to EC or not. The input u_k is quantized and takes values from $\mathcal{U} = \{\mu_1, \mu_2, \mu_3, \mu_4, \mu_5\} = \{2, 3, 8, 10.7, 19\} \subseteq \mathbf{R}$, which indicates that $\mathcal{P} = \mathcal{U}$. At k, the input sequence is supposed to generate patters with the following frequencies

$$k_1 = k - k_2 - k_3 - k_4, \quad k_2 = \lceil 0.4(k - k_3 - k_4) \rceil$$

$$k_3 = \min\{91, |\lceil \ln k \rceil|\}, \ k_4 = \lceil \sqrt{k} \rceil, \ k_5 = 2k_4$$

It follows that $\beta_1 = \lim\limits_{k\to\infty} k_1/k = 0.6, \beta_2 = 0.4, \beta_\iota = 0$ for $\iota = 3,4,5$; $\tilde{\Psi} = [2,3]^{\mathrm{T}}, [w_1, w_2]^{\mathrm{T}} = [22, 33]^{\mathrm{T}}, \bar{\alpha} = [11, 22]^{\mathrm{T}}$. Let $\pi_1 = 2$ be the basic persistent pattern. Then $\Psi = 2, \pi_2 = 1.5\pi_1, w_2 = 1.5 w_1, W = w_1 = 22, Y_1 = [1/2, 0]^{\mathrm{T}}$ and $Y_2 = [1/2, 1/18]^{\mathrm{T}}$.

Let $\mathcal{L} = I$ and $\mathbb{W} = \mathbb{W}_0 = \{[z_1, z_2]^{\mathrm{T}} \in \mathbf{R}^2 : 1 \leqslant z_1 \leqslant 10^4, |z_2| \leqslant 10^5\}$. This means that $\alpha = \bar{\alpha} \in \mathbb{W}$. For any $\varsigma = [\varsigma, \varsigma_2]^{\mathrm{T}} \in \mathbb{W}_0$, it can be derived that the equations

$$[1, \delta] Y_\iota \varepsilon = \varsigma_\iota, \quad \iota = 1, 2$$

have a unique solution $\tau_1(\varsigma) = 9(\varsigma_2/\varsigma_1 - 1)$ and $\tau_2(\varsigma) = 2\varsigma_2$, which indicates that Assumption 8.2 holds.

With the event-trigged scheme (8.4), we use (8.24) and (8.25) to compute $\hat{\eta}(k)$ and $\hat{W}(k)$. The convergence is shown by Figures 8.3 and 8.4. Then the convergence of $\hat{\theta}(k)$ from (8.26) is illustrated by Figure 8.5.

Since $\vartheta = [\eta, W]^{\mathrm{T}} = [9, 22]^{\mathrm{T}}$ and $\mathcal{H}_l(\vartheta) = [1, \eta] Y_\iota W$ for $\iota = 1, 2$, one can have

$$\frac{\partial \mathcal{H}_1(\vartheta)}{\partial \vartheta} = \begin{bmatrix} 0 \\ 1/2 \end{bmatrix}, \quad \frac{\partial \mathcal{H}_2(\vartheta)}{\partial \vartheta} = \begin{bmatrix} 11/9 \\ 1 \end{bmatrix}$$

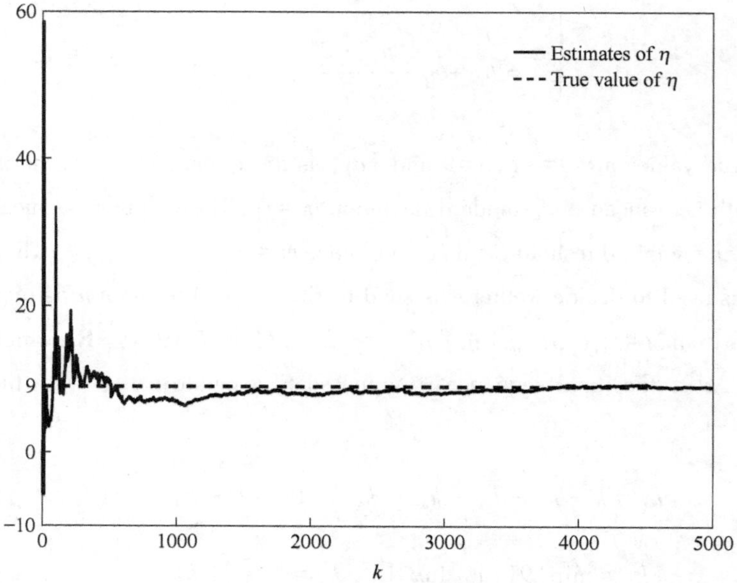

Fig. 8. 3 Convergence of $\hat{\eta}(k)$ from (8. 24)

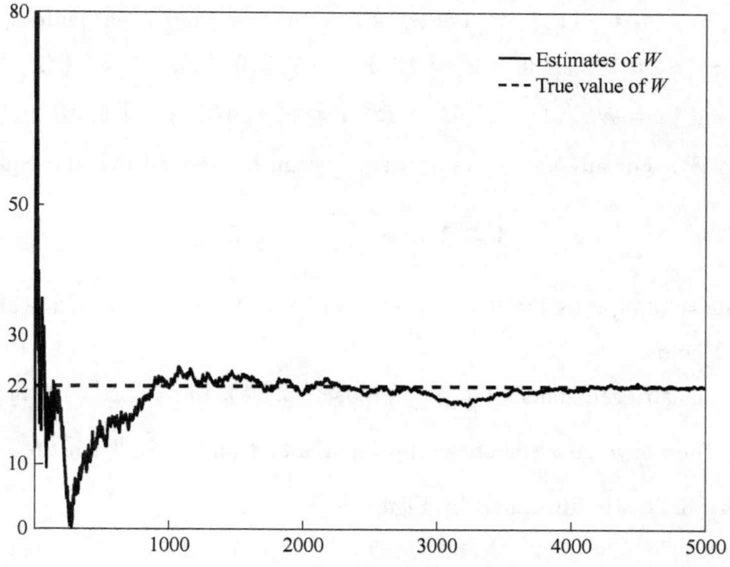

Fig. 8. 4 Convergence of $\hat{W}(k)$ from (8. 25)

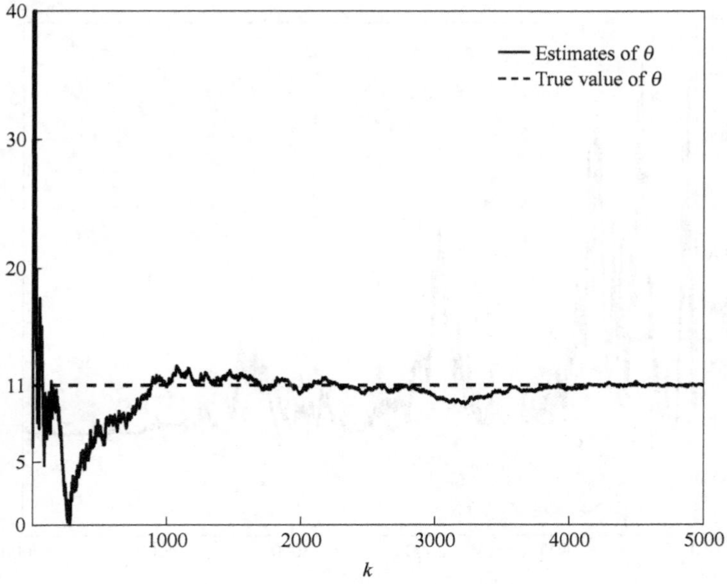

Fig. 8.5 Convergence of $\hat{\theta}(k)$ from (8.26)

By Theorem 8.3, with $\mathcal{L} = \mathcal{L}^* = \frac{1}{11}\begin{bmatrix} -18 & 22 \\ 9 & 0 \end{bmatrix}$ the asymptotic efficiency of $\hat{\vartheta}(k)$ is simulated by Figure 8.6.

The transmission on the time interval $[0,50]$ is shown in Figure 8.7, where only 4 measurements are send to the EC. In Figure 8.8, the convergence of $\frac{1}{k}\sum_{i=1}^{k}\gamma_i$ is demonstrated.

8.5 Notes

This paper studied the Hammerstein system identification with quantized inputs and quantized output observations. The problem of identifiability was discussed first, then a three-step identification algorithm was designed for the identifiable system, and finally the convergence properties were obtained, including the strong convergence, the mean-square convergence rate and the asymptotic efficiency.

Future works may be the further study with correlated noises, the robust analysis under the un-modeled dynamics, and the identification of Wiener-Hammerstein and other non-

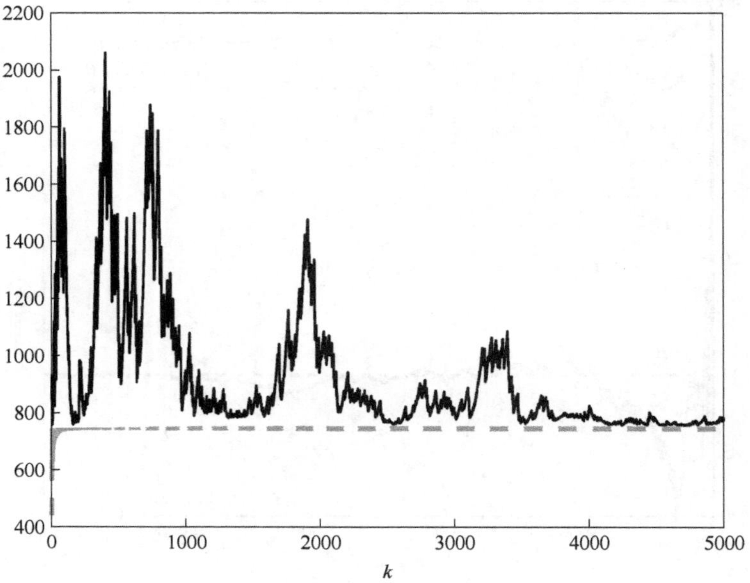

Fig. 8. 6 Asymptotic efficiency of $\hat{\vartheta}(k)$ from (8. 23) ; The dash line is the Frobenius norm of $k\sum_{CR}(k;\vartheta)$ and the solid line comes from the average of 100 trajectories of the Frobenius norm of $N(\hat{\vartheta}(k)-\vartheta)(\hat{\vartheta}(k)-\vartheta)^{\mathrm{T}}$

linear systems under quantized inputs and quantized output observations. For the further study and more details, readers are referred to the references [102~107].

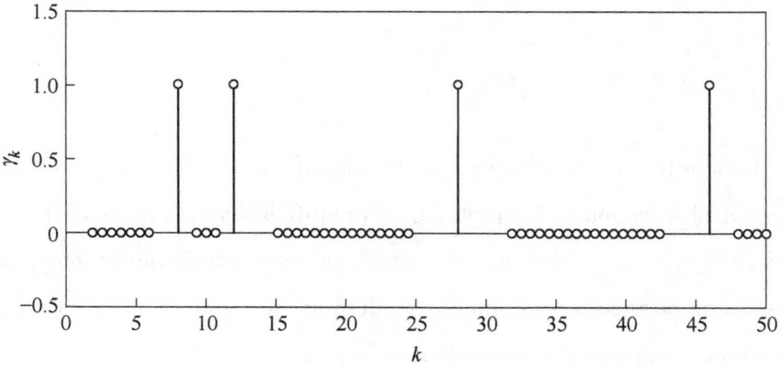

Fig. 8. 7 Transmission on the time interval [0,50]

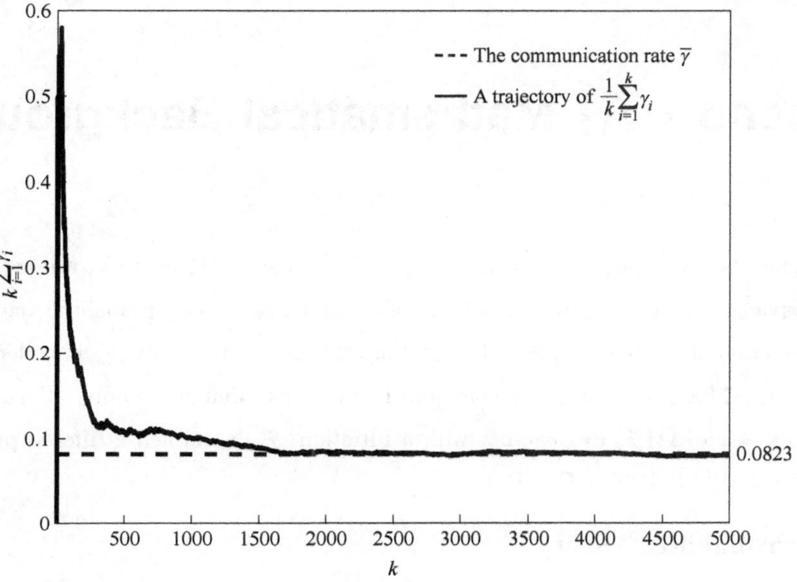

Fig. 8. 8 Communication rate

Appendix A: Mathematical Background

Throughout the book, we work with a probability space $(\Omega, \mathcal{F}, \mathrm{Pr})$, where Ω is the sample space, $\mathcal{F}$ is a σ-algebra of subsets of Ω, and $\mathrm{Pr}(\cdot)$ is a probability measure on $\mathcal{F}$. A collection of σ-algebras $\{\mathcal{F}_t\}$, for $t \geqslant 0$ or $t = 1, 2, \cdots$, or simply $\mathcal{F}_t$, is called a filtration if $\mathcal{F}_s \subset \mathcal{F}_t$ for $s \leqslant t$. The $\mathcal{F}_t$ is complete in the sense that it contains all null sets. A probability space $(\Omega, \mathcal{F}, \mathrm{Pr})$ together with a filtration $\{\mathcal{F}_t\}$ is termed a filtered probability space, denoted by $(\Omega, \mathcal{F}, \{\mathcal{F}_t\}, \mathrm{Pr})$.

A. 1 Probability Theory

A. 1. 1 Some Concepts

A. 1. 1. 1 Sample Space, Events, Probability

Consider an experiment with a number of possible outcomes. The totality of such outcomes is a sample space Ω. An event A is a subset of the sample space. A probability measure $\mathrm{Pr}(\cdot)$ is a mapping from events into the reals satisfying the axioms

(1) $\mathrm{Pr}(A) \geqslant 0$;

(2) $\mathrm{Pr}(\Omega) = 1$;

(3) For a countable set $\{A_i\}$ of events, if $A_i \cap A_j = \varnothing$ for all i, j, then $\mathrm{Pr}(\cup A_i) = \sum_i \mathrm{Pr}(A_i)$ (Here, $\varnothing$ denotes the empty set, and the set $\{A_i\}$ is termed mutually disjoint.)

Important consequences for an arbitrary countable set $\{A_i\}$ of events are

$$\mathrm{Pr}(A) \leqslant 1, \mathrm{Pr}(\varnothing) = 0, \mathrm{Pr}(\bar{A}) = 1 - \mathrm{Pr}(A), \text{and } \mathrm{Pr}(\cup A_i) \leqslant \sum_i \mathrm{Pr}(A_i)$$

with $\bar{A}$ denoting the event "not A" or "complement of A". Not all subsets of the sample space need be events, but the events must form a sigma field; in other words, if A is an event, $\bar{A}$ is an event, and if $\{A_i\}$ is a countable set of events, $\cup A_i$ is an event. Finally, Ω is an event. Frequently, it is also assumed that if A is an event with $\mathrm{Pr}(A) = 0$ and B

is any subset of A, then B is also an event with $\Pr(B)=0$. The probability measure is then termed complete.

The joint probability of two events A and B is $\Pr(A\cap B)$, sometimes written as $\Pr(AB)$.

A. 1. 1. 2 Conditional Probability

Suppose A and B are two events and an experiment is conducted with the result that event B occurs. The probability that event A has also occurred, or the conditional probability of A given B, is

$$\Pr(A|B)=\frac{\Pr(AB)}{\Pr(B)} \quad \text{assuming} \quad \Pr(B)\neq 0$$

$\Pr(A|B)$ for fixed B and variable A satisfies the probability measure axioms. (Note that the definition of $\Pr(A|B)$ when $\Pr(B)=0$ is apparently of no interest, precisely because the need to make this definition arises with zero probability.)

A. 1. 1. 3 Bayes' Rule

If $\Pr(B)\neq 0$, then

$$\Pr(A|B)=\frac{\Pr(B|A)\Pr(A)}{\Pr(B)}$$

A. 1. 1. 4 Law of Total Probability

If $\{B_n : n=1,2,\cdots\}$ is a finite or countably infinite partition of a sample space (in other words, a set of pairwise disjoint events whose union is the entire sample space) and each event B_n is measurable, then for any event A of the same probability space

$$\Pr(A)=\sum_n \Pr(A\cap B_n)$$

or, alternatively,

$$\Pr(A)=\sum_n \Pr(A|B_n)\Pr(B_n)$$

where, for any n for which $\Pr(B_n) = 0$ these terms are simply omitted from the summation, because $\Pr(A|B_n)$ is finite.

The summation can be interpreted as a weighted average, and consequently the marginal probability, $\Pr(A)$, is sometimes called "average probability"; "overall probability" is sometimes used in less formal writings.

A. 1. 1. 5 Random Variables

A random variable X is a function from the outcomes ω in a sample space Ω to the real numbers, with two properties as given below. A value of the random variable X is the number $X(\omega)$ when the outcome ω occurs. Most commonly, X can take either discrete values (X is then a discrete random variable), or continuous values in some interval $[a, b]$ (X is then a continuous random variable).

We adopt the convention that $\Pr(X=2)$ means $\Pr(\{\omega | X(\omega) = 2\})$, i. e., the probability of the subset of Ω consisting of those outcomes ω for which $X(\omega) = 2$. Likewise, $\Pr(X>0)$ means $\Pr(\{\omega | X(\omega) > 0\})$, etc.

For X to be a random variable, we require that:

(1) $\Pr(X = -\infty) = \Pr(X = +\infty) = 0$.

(2) For all real a, $\{\omega | X(\omega) \leqslant a\}$ is an event, i. e.,

$$\Pr(\{\omega | X(\omega) \leqslant a\}) = \Pr(X \leqslant a)$$

is defined.

A. 1. 1. 6 Cumulative Distribution Function

Given a random variable X, the cumulative distribution function (CDF) F_X is a mapping from the reals to the interval $[0, 1]$, i. e.,

$$F_X(x) = \Pr(X \leqslant x)$$

The subscript X on F identifies the random variable; the argument x is simply a typical value. Every CDF $F_X(\cdot)$ is non-decreasing and right-continuous. Furthermore,

$$\lim_{x \to -\infty} F_X(x) = 0, \quad \lim_{x \to +\infty} F_X(x) = 1$$

Every function with these four properties is a distribution function, i. e., for every such function, a random variable can be defined such that the function is the CDF of that random variable.

A. 1. 1. 7 Density Function

It is frequently the case that $F_X(x)$ is differentiable everywhere. The probability density function $f_X(x)$ associated with the random variable X is

$$f_X(x) = \frac{\mathrm{d}F_X(x)}{\mathrm{d}x}$$

Then $f_X(x)\,\mathrm{d}x$ to first order is $\Pr\{x < X \leqslant x + \mathrm{d}x\}$. A discrete random variable only has a density function in the sense that the density function consists of a sum of delta functions.

Example: The continuous uniform distribution or rectangular distribution is a family of symmetric probability distributions such that for each member of the family, all intervals of the same length on the distribution's support are equally probable. The support is defined by the two parameters, a and b, which are its minimum and maximum values. The distribution is often abbreviated $U(a,b)$, whose density function is

$$f(x) = \begin{cases} \dfrac{1}{b-a}, & \text{for } a \leqslant x \leqslant b \\ 0, & \text{otherwise} \end{cases}$$

as in Figure A. 1. The CDF is

$$F(x) = \begin{cases} 0, & \text{for } x < a \\ \dfrac{x-a}{b-a}, & \text{for } a \leqslant x \leqslant b \\ 1, & \text{for } x > b \end{cases}$$

as in Figure A. 2.

A. 1. 1. 8 Inverse Distribution Function

If the CDF $F(\cdot)$ is strictly increasing and continuous, then $F^{-1}(y)$, $y \in [0,1]$, is the

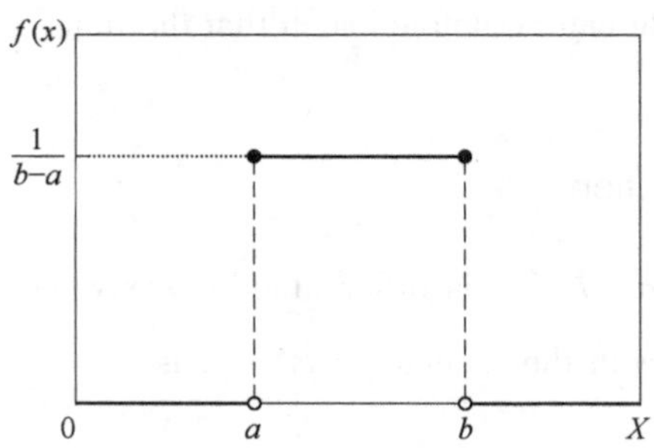

Fig. A. 1　Density function of uniform distribution $U(a,b)$

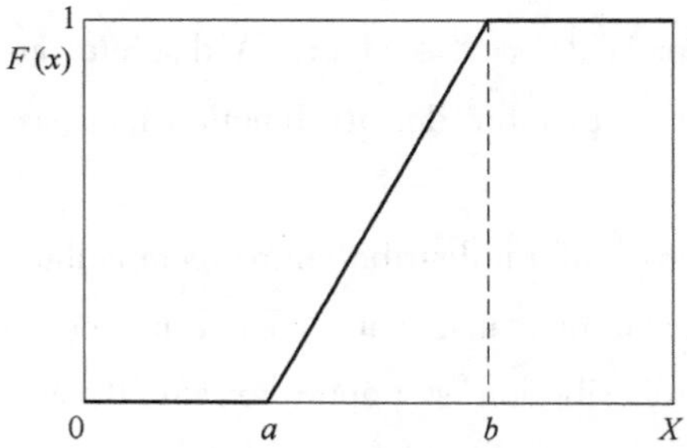

Fig. A. 2　CDF of uniform distribution $U(a,b)$

unique real number x such that $F(x)=y$. In such a case, this defines the inverse distribution function.

Some distributions do not have a unique inverse (for example in the case where $f_X(x)=0$ for all $a<x<b$, causing F_X to be constant). This problem can be solved by defining, for $y\in[0,1]$, the generalized inverse distribution function: $F^{-1}(y)=\inf\{x\in\mathbf{R}: F(x)\geqslant y\}$.

Some useful properties of the inverse CDF (which are also preserved in the definition of the generalized inverse distribution function) are

(1) F^{-1} is nondecreasing.

(2) $F^{-1}(F(x))\leqslant x$.

(3) $F(F^{-1}(y))\geqslant y$.

(4) $F^{-1}(y)\leqslant x$ if and only if $p\leqslant F(x)$.

(5) If Y has a $U[0,1]$ distribution then $F^{-1}(Y)$ is distributed as $F(\cdot)$. This is used in random number generation using the inverse transform sampling-method.

(6) If $\{X_t\}$ is a collection of independent F-distributed random variables defined on

the same sample space, then there exist random variables Y_t such that Y_t is distributed as $U[0,1]$ and $F^{-1}(Y_t)=X_t$ with probability 1 for all t.

A. 1. 1. 9 Independent Random Variables

X and Y are independent random variables if the events $\{X\leqslant x\}$ and $\{Y\leqslant y\}$ are independent for all x and y; equivalently,

$$F_{X,Y}(x,y)=F_X(x)F_Y(y)$$

or

$$f_{X,Y}(x,y)=f_X(x)f_Y(y)$$

or

$$f_{X|Y}(x|y)=f_X(x)$$

There is obvious extension to random vectors and conditional independence.

A. 1. 1. 10 Mean, Variance, and Expectation

The mean or expectation of a random variable X, written $E[X]$, is the number $\int_{-\infty}^{+\infty} xf_X(x)\,dx$, where the integral is assumed absolutely convergent. If absolute convergence does not hold, $E[X]$ is not defined. The variance σ^2 is

$$E[(X-E[X])^2]=\int_{-\infty}^{+\infty}(x-E[X])^2 f_X(x)\,dx$$

One can also show that $\sigma^2=E[X^2]-(E[X])^2$. The definition of the mean generalizes in an obvious way to a vector. For vector X, the variance is replaced by the covariance matrix

$$E\{(X-E[X])(X-E[X])^{\mathrm{T}}\}$$

The variance is always nonnegative, and the covariance matrix nonnegative definite symmetric. If $Y=g(X)$ is a function of a random variable X, the random variable Y has expected value

$$E[g(X)] = \int_{-\infty}^{+\infty} g(x) f_X(x) \mathrm{d}x$$

These notions generalize to the situation when the probability density does not exist.

A. 1. 1. 11 Normal (or Gaussian) Distribution

The normal (or Gaussian) distribution is a very common continuous probability distribution. Normal distributions are important in statistics and are often used in the natural and social sciences to represent real-valued random variables whose distributions are not known. A random variable with a Gaussian distribution is said to be normally distributed and is called a normal deviate.

The normal distribution is useful because of the central limit theorem. In its most general form, under some conditions (which include finite variance), it states that averages of samples of observations of random variables independently drawn from independent distributions converge in distribution to the normal, that is, become normally distributed when the number of observations is sufficiently large. Physical quantities that are expected to be the sum of many independent processes (such as measurement errors) often have distributions that are nearly normal. Moreover, many results and methods (such as propagation of uncertainty and least squares parameter fitting) can be derived analytically in explicit form when the relevant variables are normally distributed.

The probability density of the normal distribution is

$$f(x \mid \mu, \sigma^2) = \frac{1}{\sqrt{2\pi\sigma^2}} \mathrm{e}^{-\frac{(x-\mu)^2}{2\sigma^2}}$$

where μ is the mean or expectation of the distribution (and also its median and mode), σ is the standard deviation, and σ^2 is the variance. The simplest case of a normal distribution is known as the standard normal distribution. This is a special case when $\mu = 0$ and $\sigma = 1$.

The normal distribution is often referred to as $\mathcal{N}(\mu, \sigma^2)$. Thus when a random variable X is distributed normally with mean μ and variance σ^2, one may write $X \sim \mathcal{N}(\mu, \sigma^2)$. The density function and CDF of $\mathcal{N}(\mu, \sigma^2)$ can be seen Figure A. 3 and Figure A. 4.

A. 1. 1. 12 Law of Large Numbers

The law of large numbers is a theorem that describes the result of performing the same

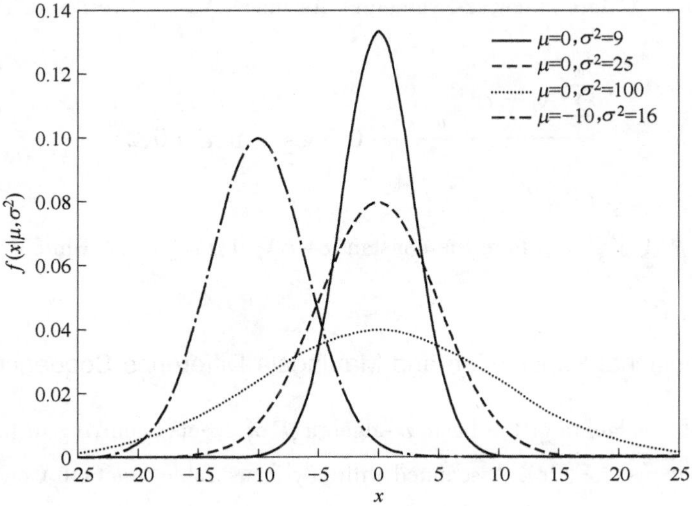

Fig. A. 3 Density function of $\mathcal{N}(\mu,\sigma^2)$

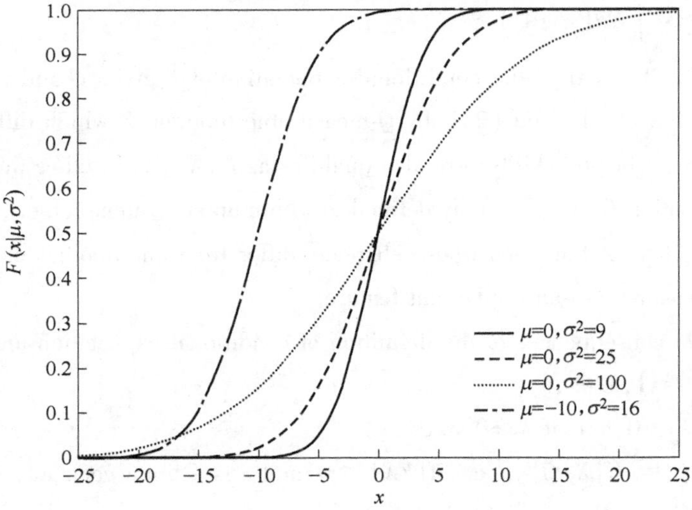

Fig. A. 4 CDF of $\mathcal{N}(\mu,\sigma^2)$

experiment a large number of times. According to the law, the average of the results obtained from a large number of trials should be close to the expected value, and will tend to become closer as more trials are performed.

A sequence of random variables $\{X_k\}_{k\geqslant1}$ is called independent and identically distributed (i. i. d.) if $\{X_k, k\geqslant1\}$ are mutually independent with the same distribution function.

Lemma A.1 (Marcinkiewicz-Zygmund) Assume $\{X_k\}_{k\geq 1}$ are i. i. d. Then

$$\frac{\sum_{k=1}^{n} X_k - cn}{n^{\frac{1}{\mu}}} \xrightarrow{n\to\infty} 0 \quad \text{a. s.} \quad \mu \in (0,2)$$

if and only if $E|X_k|^{\mu}<\infty$, where the constant $c=EX_k$ if $\mu\in[1,2)$, while c is arbitrary if $\mu\in(0,1)$.

A. 1. 2 Conditional Expectation and Martingale Difference Sequence

Let $\mathcal{G}$ be a sub-σ-algebra of the basic σ-algebra $\mathcal{F}$ of events occurring in the underlying probability space$(\Omega,\mathcal{F},\Pr)$. Associated with any measurable function X on Ω whose integral is defined, i. e., $|EX|\leqslant\infty$, is a function Y on Ω with $|EY|\leqslant\infty$ satisfying

(1) Y is $\mathcal{G}$-measurable;

(2) $\int_A Y d\Pr = \int_A X d\Pr$, all $A \in \mathcal{G}$.

Such a function Y is called the conditional expectation of X given $\mathcal{G}$ and is denoted by $E\{X|\mathcal{G}\}$. In view of (1) and (2), any $\mathcal{G}$-measurable function Z which differs from $Y=E\{X|\mathcal{G}\}$ on a set of probability zero also qualifies as $E\{X|\mathcal{G}\}$. In other words, the conditional expectation $E\{X|\mathcal{G}\}$ is only defined to within an equivalence, i. e., is any representative of a class of functions whose elements differ from one another only on sets of probability measure zero-an unpleasant feature.

Some simple consequences of the definition of conditional expectation are

(1) $E\{1|\mathcal{G}\}=1$, a. c.,

(2) $E\{X|\mathcal{G}\}\geqslant 0$, a. c. if $X\geqslant 0$, a. c.,

(3) $E\{cX|\mathcal{G}\}=cE\{X|\mathcal{G}\}$, a. c., if $|EX|\leqslant\infty$ and c is a finite constant,

(4) $E\{X+Y|\mathcal{G}\}=E\{X|\mathcal{G}\}+E\{Y|\mathcal{G}\}$, a. c.,

$$\text{if} \quad E(X^- + Y^-) < \infty \quad \text{or} \quad E(X^+ + Y^+) < \infty$$

with $Z^-=\max(-Z,0)$ and $Z^+=\max(0,Z)$. These properties assert roughly that if $\mathbb{T}X=E\{X|\mathcal{G}\}$, then $\mathbb{T}$ is linear, order preserving (monotone), and $\mathbb{T}1=1$. In addition, $E\{X|\mathcal{G}\}=X$, a. c., if X is $\mathcal{G}$-measurable.

Lemma A. 2 Let X be a random variable with $|EX| \leqslant \infty$ and $\mathcal{G}$ a σ-algebra of events. If Y is a finite-valued $\mathcal{G}$-measurable random variable such that $|EXY| \leqslant \infty$, then

$$E\{XY|\mathcal{G}\} = YE\{X|\mathcal{G}\}, \text{a. c.}$$

Definition A. 1 Let $\{X_k\}_{k\geqslant 1}$ be a sequence of random variables and $\{\mathcal{F}_k\}_{k\geqslant 1}$ be a sequence of nondecreasing σ-algebras. If X_k is $\mathcal{F}_k$-measurable for each $k\geqslant 1$, then we call $\{X_k, \mathcal{F}_k\}_{k\geqslant 1}$ an adapted process/sequence.

Lemma A. 3 Let $(X_k, \mathcal{F}_k)$, $(Y_k, \mathcal{F}_k)$ be two nonnegative adapted sequences.

(1) If $E(X_{k+1} | \mathcal{F}_k) \leqslant X_k + Y_k$ and $E\sum_{i=1}^{\infty} Y_i < \infty$, then X_k converges a. s. to a finite limit.

(2) If $E(X_{k+1} | \mathcal{F}_k) \leqslant X_k - Y_k$, then $\sum_{i=1}^{\infty} Y_i < \infty$ a. s.

An MDS (martingale difference sequence) is related to the concept of the martingale. A stochastic series X is an MDS if its expectation with respect to the past is zero. Formally, consider an adapted sequence $\{X_t, \mathcal{F}_t\}_{-\infty}^{\infty}$ on a probability space $(\Omega, \mathcal{F}, \Pr)$. X_t is an MDS if it satisfies $E|X_t| < \infty$, and $E[X_t | \mathcal{F}_{t-1}] = 0$, a. s., for all t. By construction, this implies that if Y_t is a martingale, then $X_t = Y_t - Y_{t-1}$ will be an MDS, hence the name.

Lemma A. 4 ((1) of Theorem 1. 3. 10 in [83]). Suppose that $\{X_k, \mathcal{F}_k\}$ is an MDS with

$$\sup_k E[|X_{k+1}|^{\mu_1} | \mathcal{F}_k] < \infty, \text{a. s. } \mu_1 > 2$$

and $\{g_k, \mathcal{F}_k\}$ is an adapted process. Then we have

$$\sum_{i=1}^{k} g_i X_{i+1} = O(\tilde{g}_k(\log \tilde{g}_k)^{\mu_2}), \text{a. s. } \forall\ \mu_2 > \frac{1}{2}$$

where $\tilde{g}_k = \left(\sum_{i=1}^{k} g_i^2\right)^{1/2}$.

Lemma A. 5 Consider an MDS $\{X_k, \mathcal{F}_k, k\geqslant 1\}$. If $E\left(\sum_{i=1}^{k} X_i\right)^2 < \infty$ and $\sum_{k=1}^{\infty} \frac{EX_k^2}{k^2} < \infty$, then

$$\frac{1}{k}\sum_{i=1}^{k} X_i \to 0, \text{ w. p. 1} \quad \text{as} \quad k \to \infty$$

A. 1. 3 Cramér-Rao Lower Bound

In estimation theory and statistics, the Cramér-Rao bound, Cramér-Rao lower bound (CRLB), Cramér-Rao inequality, Frechet-Darmois-Cramér-Rao inequality, or information inequality expresses a lower bound on the variance of unbiased estimators of a deterministic (fixed, though unknown) parameter. This term is named in honor of Harald Cramér, Calyampudi Radhakrishna Rao, Maurice Fréchet and Georges Darmois all of whom independently derived this limit to statistical precision in the 1940s.

In its simplest form, the bound states that the variance of any unbiased estimator is at least as high as the inverse of the Fisher information. An unbiased estimator which achieves this lower bound is said to be (fully) efficient. Such a solution achieves the lowest possible mean squared error among all unbiased methods, and is therefore the minimum variance unbiased (MVU) estimator. However, in some cases, no unbiased technique exists which achieves the bound. This may occur even when an MVU estimator exists.

Suppose θ is an unknown deterministic parameter which is to be estimated from measurements x, distributed according to some probability density function $f(x;\theta)$. The variance of any unbiased estimator $\hat{\theta}$ of θ is then bounded by the reciprocal of the Fisher information $I(\theta)$:

$$\operatorname{var}(\hat{\theta}) \geqslant \frac{1}{I(\theta)}$$

where the Fisher information $I(\theta)$ is defined by

$$I(\theta) = E\left[\left(\frac{\partial l(x;\theta)}{\partial \theta}\right)^2\right] = -E\left[\frac{\partial^2 l(x;\theta)}{\partial \theta^2}\right]$$

and $l(x;\theta) = \log(f(x;\theta))$ is the natural logarithm of the likelihood function.

A more general form of the bound can be obtained by considering a biased estimator $T(X)$ of a function of the parameter, $\psi(\theta)$. Here, biasedness is understood as stating that $E\{T(X)\} = \psi(\theta)$ is not generally equal to 0. In this case, the bound is given by

$$\operatorname{var}(T) \geqslant \frac{[\psi'(\theta)]^2}{I(\theta)}$$

where $\psi'(\theta)$ is the derivative of $\psi(\theta)$ (by θ), and $I(\theta)$ is the Fisher information defined above.

Extending the CRLB to multiple parameters, define a parameter column vector

$$\theta = [\theta_1, \cdots, \theta_d]^{\mathrm{T}} \in R^d$$

with probability density function $f(x;\theta)$ which satisfies the two conditions below.

(1) The Fisher information is always defined; equivalently, for all x such that $f(x;\theta) > 0$,

$$\frac{\partial}{\partial \theta} \log f(x;\theta)$$

exists, and is finite.

(2) The operations of integration with respect to x and differentiation with respect to θ can be interchanged in the expectation of T; that is,

$$\frac{\partial}{\partial \theta}\left[\int T(x) f(x;\theta)\,\mathrm{d}x\right] = \int T(x)\left[\frac{\partial}{\partial \theta} f(x;\theta)\right]\mathrm{d}x$$

whenever the right-hand side is finite.

The Fisher information matrix is a $d \times d$ matrix with element $I_{m,k}$ defined as

$$I_{m,k} = E\left[\frac{\partial}{\partial \theta_m}\log f(x;\theta)\frac{\partial}{\partial \theta_k}\log f(x;\theta)\right] = -E\left[\frac{\partial^2}{\partial \theta_m \partial \theta_k}\log f(x;\theta)\right]$$

Let $T(X)$ be an estimator of any vector function of parameters, $T(X) = (T_1(X), \cdots, T_d(X))^{\mathrm{T}}$, and denote its expectation vector $E[T(X)]$ by $\psi(\theta)$. The CRLB then states that the covariance matrix of $T(X)$ satisfies

$$\operatorname{cov}_\theta(T(X)) \geqslant \frac{\partial \psi(\theta)}{\partial \theta}[I(\theta)]^{-1}\left(\frac{\partial \psi(\theta)}{\partial \theta}\right)^{\mathrm{T}}$$

where

—The matrix inequality $A \geqslant B$ is understood to mean that the matrix $A - B$ is positive semidefinite, and

—$\partial\psi(\theta)/\partial\theta$ is the Jacobian matrix whose ij element is given by $\partial\psi_i(\theta)/\partial\theta_j$.

If $T(X)$ is an unbiased estimator of θ(i. e. ,$\psi(\theta)=\theta$) ,then the CRLB reduces to

$$\mathrm{cov}_\theta(T(X)) \geqslant I(\theta)^{-1}$$

If it is inconvenient to compute the inverse of the Fisher information matrix, then one can simply take the reciprocal of the corresponding diagonal element to find a (possibly loose) lower bound.

$$\mathrm{var}_\theta(T_m(X)) = [\mathrm{cov}_\theta(T(X))]_{mm} \geqslant [I(\theta)^{-1}]_{mm} \geqslant ([I(\theta)]_{mm})^{-1}$$

Suppose, in addition, that the operations of integration and differentiation can be swapped for the second derivative of $f(x;\theta)$ as well, i. e. ,

$$\frac{\partial^2}{\partial\theta^2}\left[\int T(x)f(x;\theta)\,\mathrm{d}x\right] = \int T(x)\left[\frac{\partial^2}{\partial\theta^2}f(x;\theta)\right]\mathrm{d}x$$

In this case, it can be shown that the Fisher information equals

$$I(\theta) = -E\left[\frac{\partial^2}{\partial\theta^2}\log f(X;\theta)\right]$$

The CRLB can then be written as

$$\mathrm{var}(\hat{\theta}) \geqslant \frac{1}{I(\theta)} = \frac{1}{-E\left[\frac{\partial^2}{\partial\theta^2}\log f(X;\theta)\right]}$$

In some cases, this formula gives a more convenient technique for evaluating the bound.

For two unbiased estimators $\hat{\theta}_i$ of θ, with $E\hat{\theta}_i^2<\infty$ $(i=1,2)$, we say that $\hat{\theta}_1$ is more efficient than $\hat{\theta}_2$ if the relative efficiency $\mathrm{eff}_\theta(\hat{\theta}_1 \mid \hat{\theta}_2) < 1$, where

$$\mathrm{eff}_\theta(\hat{\theta}_1 \mid \hat{\theta}_2) = \frac{\mathrm{var}_\theta(\hat{\theta}_1)}{\mathrm{var}_\theta(\hat{\theta}_2)}$$

We say that an estimator $\hat{\theta}$ is efficient or most efficient if the CRLB is attained. For any unbiased estimator $\hat{\theta}_1$, the efficiency of the estimator is defined as $\mathrm{eff}_\theta(\hat{\theta}_1 \mid \hat{\theta})$, where $\hat{\theta}$

is an efficient estimator. An estimator $\hat{\theta}_1$ with sample size N is asymptotically efficient if

(1) $\hat{\theta}_1$ is at least asymptotically unbiased in the sense that $E\hat{\theta}_1 \to \theta$ as $N \to \infty$, and

(2) $\lim_{N} \mathrm{eff}_\theta(\hat{\theta}_1 \mid \hat{\theta}) = 1$.

A.2 Vector and Matrix

A.2.1 Vector Norm

A norm is a function that assigns a strictly positive length or size to each vector in a vector spacesave for the zero vector, which is assigned a length of zero. A vector space on which a norm is defined is called a normed vector space.

Given a vector space V over a subfield K of the complex numbers, a norm on V is a nonnegative-valued scalar function $\mathcal{T}: V \to [0, +\infty)$ with the following properties:

For all $a \in K$ and all $u, v \in V$

(1) $\mathcal{T}(u+v) \leqslant \mathcal{T}(u) + \mathcal{T}(v)$ (being subadditive or satisfying the triangle inequality).

(2) $\mathcal{T}(av) = |a| \mathcal{T}(v)$ (being absolutely homogeneous or absolutely scalable).

(3) If $\mathcal{T}(v) = 0$, then $v = 0$ is the zero vector (being positive definite or being point-separating).

Two norms $\mathcal{T}$ and $\mathcal{S}$ on a vector space V are equivalent if there exist two real constants c_1 and c_2, with $c_1 > 0$, such that for every vector v in V, one has that:

$$c_1 \mathcal{S}(v) \leqslant \mathcal{T}(v) \leqslant c_2 \mathcal{S}(v)$$

If a norm $\mathcal{T}: V \to [0, +\infty)$ is given on a vector space V, then the norm of a vector $v \in V$ is usually denoted by enclosing it within double vertical lines: $\| v \| = \mathcal{T}(v)$.

On an n-dimensional space $\mathbf{R}^n$, the intuitive notion of length of the vector $\boldsymbol{x} = (x_1, \cdots, x_n)^{\mathrm{T}}$ is captured by the formula

$$\| x \| := \sqrt{x_1^2 + \cdots + x_n^2}$$

This gives the ordinary distance from the origin to the point x, and is called Euclidean norm.

A.2.2 Matrix Norm

A matrix norm on the vector space $\mathbf{R}^{n\times n}$ is a function $\| \cdot \| : \mathbf{R}^{n\times n} \to \mathbf{R}$ that must satisfy the following properties:

For all scalars $a \in \mathbf{R}$ and for all matrices A and B in $\mathbf{R}^{n \times n}$,

(1) $\| aA \| = |a| \| A \|$ (being absolutely homogeneous);

(2) $\| A+B \| \leqslant \| A \| + \| B \|$ (being sub-additive or satisfying the triangle inequality);

(3) $\| A \| \geqslant 0$ (being positive-valued), and $\| A \| = 0$ if and only if $A = 0_{m \times n}$ (being definite);

(4) $\| AB \| \leqslant \| A \| \| B \|$ (being sub-multiplicative).

The Frobenius norm can be defined in various ways:

$$\| A \| = \sqrt{\sum_{i=1}^{n} \sum_{j=1}^{n} |a_{ij}|^2} = \sqrt{\mathrm{trace}(A^{\mathrm{T}} A)} = \sqrt{\sum_{i=1}^{n} \sigma_i^2(A)}$$

where $\sigma_i(A)$ are the singular values of A.

A. 2. 3 Moore-Penrose Inverse

A pseudoinverse A^+ of a matrix A is a generalization of the inverse matrix. The most widely known type of matrix pseudoinverse is the Moore-Penrose inverse, which was independently described by E. H. Moore in 1920, Arne Bjerhammar in 1951, and Roger Penrose in 1955. Earlier, Erik Ivar Fredholm had introduced the concept of a pseudoinverse of integral operators in 1903. When referring to a matrix, the term pseudoinverse, without further specification, is often used to indicate the Moore-Penrose inverse. The term generalized inverse is sometimes used as a synonym for pseudoinverse.

The pseudoinverse is defined and unique for all matrices whose entries are real or complex numbers. It can be computed using the singular value decomposition.

A pseudoinverse of A is defined as a matrix A^+ satisfying all of the following four criteria, known as the Moore-Penrose conditions:

(1) $AA^+A = A$ (AA^+ need not be the general identity matrix, but it maps all column vectors of A to themselves);

(2) $A^+AA^+ = A^+$ (A^+ is a weak inverse for the multiplicative semigroup);

(3) $(AA^+)^* = AA^+$ (AA^+ is Hermitian);

(4) $(A^+A)^* = A^+A$ (A^+A is also Hermitian).

where A^* denotes the Hermitian/conjugate transpose.

A^+ exists for any matrix A, but when the latter has full rank, A^+ can be expressed as a simple algebraic formula. In particular, when A has linearly independent columns (and thus matrix A^*A is invertible), A^+ can be computed as:

$$A^+ = (A^*A)^{-1}A^*$$

This particular pseudoinverse constitutes a left inverse, since, in this case, $A^+A = I$. When A has linearly independent rows (matrix AA^* is invertible), A^+ can be computed as:

$$A^+ = A^*(AA^*)^{-1}$$

This is a right inverse, as $AA^+ = I$.

References

[1] Zadeh L A. On the identification problem [J]. IEEE Transactions on Circuits and Systems, 1956, 3 (4): 277~281.

[2] Söderström T, Stoica P. System Identification [M]. Prentice Hall, Upper Saddle River, N. J, 1989.

[3] Aström K J, Bernhardsson B M. Comparison of Riemann and Lebesgue sampling for first order stochastic systems [C]. In Proceedings of the 41st IEEE Conference on Decision and Control, Las Vegas, Nevada, USA, 2002, 2: 2011~2016.

[4] Lemmon M. Event-triggered feedback in control, estimation, and optimization [J]. Lecture Notes in Control and Information Sciences, 2010, 406: 293~358.

[5] Mahmoud M S, Sabih M. Networked event-triggered control: an introduction and research trends [J]. International Journal of General Systems, 2014, 43(8): 810~827.

[6] Han D, Mo Y, Wu J, Weerakkody S, Sinopoli B, Shi L. Stochastic event-triggered sensor schedule for remote state estimation [J]. IEEE Transactions on Automatic Control, 2015, 60(10): 2661~2675.

[7] Dimarogonas D V, Frazzoli E, Johansson K H. Distributed event-triggered control for multi-agent systems [J]. IEEE Transactions on Automatic Control, 2012, 57(5): 1291~1297.

[8] Fan Q Y, Yang G H. Event-based fuzzy adaptive fault-tolerant control for a class of nonlinear systems [J]. IEEE Transactions on Fuzzy Systems, 2018, DOI: 10.1109/TFUZZ.2018.2800724.

[9] Peng C, Yang T C. Event-triggered communication and H_∞ control co-design for networked control systems [J]. Automatica, 2013, 49(5): 1326~1332.

[10] Weerakkody S, Mo Y, Sinopoli B, Han D, Shi L. Multi-sensor scheduling for state estimation with event-based, stochastic triggers [J]. IEEE Transactions on Automatic Control, 2016, 61(9): 2695~2701.

[11] Zhang X M, Han Q L. Event-based H_∞ filtering for sampled-data systems [J]. Automatica, 2015, 51: 55~69.

[12] Tabuada P. Event-triggered real-time scheduling of stabilizing control tasks [J]. IEEE Transactions on Automatic Control, 2007, 52(9): 1680~1685.

[13] Förstner D, Lunze J. Discrete-event models of quantized systems for diagnosis [J]. International Journal of Control, 2010, 74(7): 690~700.

[14] Årzén K E. A simple event-based PID controller [C]. In Proceedings of the 14th IFAC world congress, Beijing, China, 1999.

[15] Fiter C, Omran H, Seuret A, Fridman E, Richarda J P, Niculescu S. Recent developments on the stability of systems with aperiodic sampling: an overview [J]. Automatica, 2017, 76: 309~335.

[16] Wang L Y, Zhang J F, Yin G G. System identification using binary sensors [J]. IEEE Transactions on Automatic Control, 2003, 48: 1892~1907.

[17] Wang L Y, Yin G G, Zhang J F, Zhao Y L. System Identification with Quantized Observations

[M]. Birkhäuser Boston,2010.

[18] Casini M, Garulli A, Vicino A. Input design in worst-case system identification with quantized measurements [J]. Automatica,2012,48(12): 2997~3007.

[19] Colinet E,Juillard J. A weighted least-squares approach to parameter estimation problems based on binary measurements [J]. IEEE Transactions on Automatic Control,2010,55(1): 148~152.

[20] Goudjil A,Pouliquen M,Pigeon E. Identification of systems using binary sensors via Support Vector Machines [C]. In Proceedings of IEEE 54th Annual Conference on Decision and Control,2015.

[21] Gustafsson F,Karlsson R. Statistical results for system identification based on quantized observations [J]. Automatica,2009,45(12): 2794~2801.

[22] Wang L Y,Zhao W. System identification: new paradigms, challenges, and opportunities [J]. Acta Automatica Sinica,2013,39(7): 933~942.

[23] Kang G,Bi W,Zhao Y,Zhang J F,Yang J,Xu H,Loh M L,Hunger S P,Relling M V,Pounds S, Cheng C. A new system identification approach to identify genetic variants in sequencing studies for a binary phenotype [J]. Human Heredity,2014,78(2): 104~116.

[24] Wang T,Bi W,Zhao Y,Xue W. Radar target recognition algorithm based on RCS observation sequence — set-valued identification method [J]. Journal of Systems Science and Complexity, 2016,29(3): 573~588.

[25] Akyildiz I,Su W,Sankarasubramaniam Y,Cayirci E. Wireless sensor networks: a survey [J]. Computer Networks,2002,38: 393~422.

[26] Casini M, Garulli A, Vicino A. Input design in worst-case system identification using binary sensors [J]. IEEE Transactions on Automatic Control,2011,56(5): 1186~1191.

[27] Cerone V,Piga D,Regruto D. Fixed-order FIR approximation of linear systems from quantized input and output data [J]. Systems & Control Letters,2013,62(12): 1136~1142.

[28] Zhao Y,Wang T,Bi W. Consensus protocol for multi-agent systems with undirected topologies and binary-valued communications [J]. IEEE Transactions on Automatic Control, 2018, DOI: 10. 1109/TAC. 2018. 2814632.

[29] Bottegal G, Hjalmarsson H, Pillonetto G. A new kernel-based approach to system identification with quantized output data [J]. Automatica,2017,85(12): 145~152.

[30] Wigren T. Adaptive filtering using quantized output measurements [J]. IEEE Transactions on Signal Processing,1998,46(12): 3423~3426.

[31] Wang J,Zhang Q. Identification of FIR systems based on quantized output measurements: a quadratic programming-based method [J]. IEEE Transactions on Automatic Control,2015,60(5): 1439~1444.

[32] Hespanha J P,Naghshtabrizi P,Xu Y. A survey of recent results in networked control systems [J]. Proceedings of the IEEE,2007,95(1): 138~162.

[33] Gupta R A,Chow M Y. Networked control system: overview and research trends [J]. IEEE

Transactions on Industrial Electronics,2010,57(7): 2527~2535.

[34] Kim K D,Kumar P R. Cyber-physical systems: a perspective at the centennial [J]. Proceedings of the IEEE,2012,100: 1287~1308.

[35] Khaitan S K,Mccalley J D. Design techniques and applications of cyberphysical systems: a survey [J]. IEEE Systems Journal,2015,9(2): 350~365.

[36] Sycara K P. Multi-agent systems [J]. AI Magazine,1998,19(2): 79~92.

[37] Chen H F. Stochastic Approximation & Its Application [M]. Kluwer Academic Publishers,Dordrecht,2002.

[38] Kushner H J, Yin G G. Stochastic Approximation and Recursive Algorithms and Applications [M]. Springer-Verlag New York,Inc,2003.

[39] Wang L Y,Yin G G. Asymptotically efficient parameter estimation using quantized output observations [J]. Automatica,2007,43: 1178~1191.

[40] Mei H W,Yin G G,Wang L Y. Almost sure convergence rates for system identification using binary,quantized,and regular sensors [J]. Automatica,2014,50(8): 2120~2127.

[41] Guo J, Wang L Y, Yin G G, Zhao Y L, Zhang J F. Asymptotically efficient identification of FIR systems with quantized observations and general quantized inputs [J]. Automatica, 2015,57: 113~122.

[42] Zhao Y L,Wang L Y,Yin G G,Zhang J F. Identification of Wiener systems with binary-valued output observations [J]. Automatica,2007,43(10): 1752~1765.

[43] Zhao Y L,Wang L Y,Yin G G,Zhang J F. Identification of Hammerstein systems with quantized observations [J]. SIAM Journal on Control and Optimization,2010,48(7): 4352~4376.

[44] Wang L Y,Yin G G,Zhang J F. Joint identification of plant rational models and noise distribution functions using binary-valued observations [J]. Automatica,2006,42: 535~547.

[45] Yin G G,Wang L Y,Kan S. Tracking and identification of regime-switching systems using binary sensors [J]. Automatica,2009,45: 944~955.

[46] Guo J,Wang L Y,Yin G G,Zhao Y L,Zhang J F. Identification of Wiener systems with quantized inputs and binary-valued output observations [J]. Automatica,2017,78: 280~286.

[47] Guo J,Liu H. Hammerstein system identification with quantized inputs and quantized output observations [J]. IET Control Theory & Applications,2017,11(4): 593~599.

[48] Guo J,Diao J D. Parameter estimation for systems with structural uncertainties based on quantized inputs and binary-valued output observations [J]. IET Control Theory & Applications,2018,12(5): 694~699.

[49] He Q,Wang L Y,Yin G G. System Identification Using Regular and Quantized Observations: Applications of Large Deviations Principles [M]. SpringerBriefs in Mathematics, New York, Springer,2013.

[50] Ding K,Li Y,Quevedo D E,Dey S,Shi L. A multi-channel transmission schedule for remote state estimation under DoS attacks [J]. Automatica,2017,78: 194~201.

[51] Gao X, Akyol E, Başar T. Optimal communication scheduling and remote estimation over an additive noise channel [J]. Automatica, 2018, 88: 57~69.

[52] Mamduhi M H, Molin A, Tolić D, Hirche S. Error-dependent data scheduling in resource-aware multi-loop networked control systems [J]. Automatica, 2017, 81: 209~216.

[53] Leong A S, Dey S, Quevedo D E. Transmission scheduling for remote state estimation and control with an energy harvesting sensor [J]. Automatica, 2018, 91: 54~60.

[54] Wang T, Zhou C, Lu H, He J, Guo J. Hybrid scheduling and quantized output feedback control for networked control systems [J]. International Journal of Control, Automation & Systems, 2018, 16 (1): 197~206.

[55] Fellouris G. Asymptotically optimal parameter estimation under communication constraints [J]. Annals of Statistics, 2012, 40(4): 2239~2265.

[56] You K, Xie L, Song S. Asymptotically optimal parameter estimation with scheduled measurements [J]. IEEE Transactions on Signal Processing, 2013, 61(14): 3521~3531.

[57] Han D, You K, Xie L, Wu J, Shi L. Optimal parameter estimation under controlled communication over sensor networks [J]. IEEE Transactions on Signal Processing, 2015, 63(24): 6473~6485.

[58] Han D, You K, Xie L, Wu J, Shi L. Stochastic packet scheduling for optimal parameter estimation [C]. IEEE Conference on Decision and Control, 2016, 3057~3062.

[59] Tiberi U, Fischione C, Johansson K H, Benedetto M D. Energy-efficient sampling of networked control systems over IEEE 802. 15. 4 wireless networks [J]. Automatica, 2013, 49(3): 712~724.

[60] Garcia E, Antsaklis P J. Model-based event-triggered control for systems with quantization and time-varying network delays [J]. IEEE Transactions on Automatic Control, 2013, 58(2): 422~434.

[61] Dolk V S, Borgers D P, Heemels W P M H. Output-based and decentralized dynamic event-triggered control with guaranteed $\mathcal{L}_p$-gain performance and zeno-freeness [J]. IEEE Transactions on Automatic Control, 2016, 62(1): 34~49.

[62] Abdelrahim M, Postoyan R, Daafouz J, Nešić D. Robust event-triggered output feedback controllers for nonlinear systems [J]. Automatica, 2017, 75: 96~108.

[63] Dong L, Zhong X, Sun C, He H. Adaptive event-triggered control based on heuristic dynamic programming for nonlinear discrete-time systems [J]. IEEE Transaction on Neural Networks and Learning Systems, 2017, 28(7): 1594~1605.

[64] Shi D, Chen T, Shi L. An event-triggered approach to state estimation with multiple point- and set-valued measurements [J]. Automatica, 2014, 50(6): 1641~1648.

[65] Shi D, Chen T, Shi L. On set-valued Kalman filtering and its application to event-based state estimation [J]. IEEE Transactions on Automatic Control, 2015, 60(5): 1275~1290.

[66] Tanwani A, Prieur C, Fiacchini M. Observer-based feedback stabilization of linear systems with event-triggered sampling and dynamic quantization [J]. Systems & Control Letters,

2016,94: 46~56.

[67] Du D,Qi B,Fei M,Wang Z. Quantized control of distributed event-triggered networked control systems with hybrid wired-wireless networks communication constraints [J]. Information Sciences,2017,380: 74~91.

[68] Godoy B I,Goodwin G C,Agüero J C,Marelli D,Wigren T. On identification of FIR systems having quantized output data [J]. Automatica,2011,47(9): 1905~1915.

[69] Xia Y,Fu M,Liu G P. Analysis and Synthesis of Networked Control Systems [M]. Springer-Verlag,Berlin Heidelberg,2011.

[70] Zhang W,Branicky M S,Phillips S M. Stability of networked control systems [J]. IEEE Control Systems Magazine,2001,21(1): 84~99.

[71] Garcia E,Cao Y,Casbeer D W. Periodic event-triggered synchronization of linear multi-agent systems with communication delays [J]. IEEE Transactions on Automatic Control,2017,62(1): 366~371.

[72] Yu H,Antsaklis P J. Event-triggered output feedback control for networked control systems using passivity achieving stability in the presence of communication delays and signal quantization [J]. Automatica,2013,49(1): 30~38.

[73] Heemels W P M H,Sandee J H,Van Den Bosch P P J. Analysis of event-driven controllers for linear systems [J]. International Journal of Control,2008,81(4): 571~590.

[74] Marellfi D,You K,Fu M. Identification of ARMA models using intermittent and quantized output observations [J]. Automatica,2013,49(2): 360~369.

[75] Godoy B I, Agüero J C, Carvajal R, Goodwin G C, Yuz J I. Identification of sparse FIR systems using a general quantisation scheme [J]. International Journal of Control,2014,87(4): 874~886.

[76] Zhao Y L,Bi W J,Wang T. Iterative parameter estimate with batched binary-valued observations [J]. SCIENCE CHINA Information Sciences,2016,59(5): 1~18.

[77] Guo J,Zhao Y L. Recursive projection algorithm on FIR system identification with binary-valued observations [J]. Automatica,2013,49: 3396~3401.

[78] You K. Recursive algorithms for parameter estimation with adaptive quantizer [J]. Automatica,2015,52: 192~201.

[79] Zhao W,Chen H F,Tempo R,Dabbene F. Recursive nonparametric identification of nonlinear systems with adaptive binary sensors [J]. IEEE Transactions on Automatic Control,2017,62(8): 3959~3971.

[80] Calamai P H,Moré J J. Projected gradient methods for linearly constrained problems [J]. Mathematical Programming,1987,39(1): 93~116.

[81] Chow Y S,Teicher H. Probability Theory: Independence,Interchangeability,Martingales [M]. Springer-Verlag,New York,2rd Edition,1997.

[82] Guo L. Time-Varying Stochastic Systems—Stability,Estimation and Control [M]. Jilin Science

and Technology Press, Changchun, P. R. China, 1993.

[83] Ribeiro A, Giannakis G B, Roumeliotis S I. SOI-KF: Distributed Kalman filtering with low-cost communications using the sign of innovations [J]. IEEE Transactions on Signal Processing, 2006, 54(12): 4782~4795.

[84] Wu J, Jia Q S, Johansson K H, Shi L. Event-based sensor data scheduling: trade-off between communication rate and estimation quality [J]. IEEE Transactions on Automatic Control, 2013, 58(4): 1041~1046.

[85] Shi D, Chen T, Shi L. Event-triggered maximum likelihood state estimation [J]. Automatica, 2014, 50(1): 247~254.

[86] Wiener N. Nonlinear Problems in Random Theory [M]. Wiley, New York, 1958.

[87] Zhu Y. Distillation column identification for control using Wiener model [C]. In Proceeding of 1999 American Control Conference, Hyatt Regency San Diego, California, USA, 1999.

[88] Kalafatis A, Arifin N, Wang L, Cluett W R. A new approach to the identification of pH processes based on the Wiener model [J]. Chemical Engineering Science, 1995, 50(23): 3693~3701.

[89] Hunter I W, Korenberg M J. The identification of nonlinear biological systems: Wiener and Hammerstein cascade models [J]. Biological Cybernetics, 1986, 55: 135~144.

[90] Boyd S, Chua L O. Fading memory and the problem of approximating nonlinear operators with Volterra series [J]. IEEE Transactions on Circuits and Systems, 1985, 32: 1150~1161.

[91] Ljung L. System Identification: Theory for the User [M]. Englewood Cliffs, NJ: Prentice-Hall, 1987.

[92] Kang Y, Zhai D H, Liu G P, Zhao Y B. On input-to-state stability of switched stochastic nonlinear systems under extended asynchronous switching [J]. IEEE Transaction on Cybernetics, 2015, 46(5): 1092~1105.

[93] He W, Zhang S, Ge S S. Adaptive control of a flexible crane system with the boundary output constraint [J]. IEEE Transactions on Industrial Electronics, 2014, 61(8): 4126~4133.

[94] Magnus J R, Neudecker H. Matrix Differential Calculus with Applications in Statistics and Econometrics [M]. John Wiley & Sons, 3rd Edition, 2007.

[95] Hagenblad A, Ljung L, Wills A. Maximum likelihood identification of Wiener models [J]. Automatica, 2008, 44(11): 2697~2705.

[96] Wang D, Ding F. Least squares based and gradient based iterative identification for Wiener nonlinear systems [J]. Signal Processing, 2011, 91(5): 1182~1189.

[97] Chen H, Zhao W. Recursive Identification and Parameter Estimation [M]. CRC Press, 2014.

[98] Greblicki W. Nonparametric approach to Wiener system identification [J]. IEEE Transactions on Circuits and Systems- Ⅰ: Fundamental Theory and Applications, 1997, 44(6): 538~545.

[99] Wills A, Schon T, Ljung L, Ninness B. Identification of Hammerstein-Wiener models [J]. Automatica, 2011, 1(1): 1~14.

[100] Giri F, Bai E W. Block-oriented Nonlinear System Identification [M]. Springer, 2010.

[101] Wang L Y, Yin G G. Quantized identification with dependent noise and Fisher information ratio of communication channels [J]. IEEE Transactions on Automatic Control, 2010, 55 (3): 674~690.

[102] Bottegal G, Pillonetto G, Hjalmarsson H. Bayesian kernel-based system identification with quantized output data [J]. Statistics, 2015, 48(28): 455~460.

[103] Zhao W, Chen H F, Tempo R, Dabbene F. Recursive identification of nonparametric nonlinear systems with binary-valued output observations [C]. In Proceedings of the 54th IEEE Conference on Decision and Control, 2015, 121~126.

[104] Narendra K, Gallman P. An iterative method for the identification of nonlinear systems using a Hammerstein model [J]. IEEE Transactions on Automatic Control, 1966, 11: 546~550.

[105] Ninness B, Gibson S. Quantifying the accuracy of Hammerstein model estimation [J]. Automatica, 2002, 38: 2037~2051.

[106] Liu Y, Bai E. Iterative identification of Hammerstein systems [J]. Automatica, 2007, 43(2): 346~354.

[107] Zhao W, Chen H F. Recursive identification for Hammerstein system with ARX subsystem [J]. IEEE Transactions on Automatic Control, 2007, 51(12): 1966~1974.